Sustainable Development Goals Series

The **Sustainable Development Goals Series** is Springer Nature's inaugural cross-imprint book series that addresses and supports the United Nations' seventeen Sustainable Development Goals. The series fosters comprehensive research focused on these global targets and endeavours to address some of society's greatest grand challenges. The SDGs are inherently multidisciplinary, and they bring people working across different fields together and working towards a common goal. In this spirit, the Sustainable Development Goals series is the first at Springer Nature to publish books under both the Springer and Palgrave Macmillan imprints, bringing the strengths of our imprints together.

The Sustainable Development Goals Series is organized into eighteen subseries: one subseries based around each of the seventeen respective Sustainable Development Goals, and an eighteenth subseries, "Connecting the Goals", which serves as a home for volumes addressing multiple goals or studying the SDGs as a whole. Each subseries is guided by an expert Subseries Advisor with years or decades of experience studying and addressing core components of their respective Goal.

The SDG Series has a remit as broad as the SDGs themselves, and contributions are welcome from scientists, academics, policymakers, and researchers working in fields related to any of the seventeen goals. If you are interested in contributing a monograph or curated volume to the series, please contact the Publishers: Zachary Romano [Springer; zachary.romano@springer.com] and Rachael Ballard [Palgrave Macmillan; rachael.ballard@palgrave.com].

Ebes Aziegbe-Esho

On the Sustainable Development of African Countries

A Strategic Human Capital Approach

Ebes Aziegbe-Esho
University of Johannesburg
Johannesburg, South Africa

Nigerian University of Technology
and Management (NUTM)
Lagos, Nigeria

ISSN 2523-3084   ISSN 2523-3092   (electronic)
Sustainable Development Goals Series
ISBN 978-3-031-81123-4   ISBN 978-3-031-81124-1 (eBook)
https://doi.org/10.1007/978-3-031-81124-1

To my parents, Engr. Daniel Ighalo Aziegbe and Roseline Aziegbe; thank you for your sacrifices. Daddy, thank you for teaching me at home every day when I returned from school.

Acknowledgments

I acknowledge everyone that has helped in one way or the other on this research journey into human capital, and how it creates value. So, many thanks to Dr. Tunji Adegbesan, my doctoral research supervisor, Professor Ifedapo Adeleye, Professor Ogechi Adeola, and Professor Grietjie Verhoef—for all the opportunities you offered me to learn.

Contents

About the Author

Ebes Aziegbe-Esho is a Senior Research Associate at University of Johannesburg (UJ), South Africa, and faculty member at Nigerian University of Management and Technology (NUTM). She completed her PhD in Management, specializing in business strategy, at Lagos Business School, Pan-Atlantic University, Nigeria (LBS-PAU). In 2016, her research paper on human capital won the best doctoral paper award at the Academy of International Business (AIB-SSA, now AIB-Africa) conference. With a focus on African context, her core research interests include exploring the creation of value from human capital resources, Foreign Direct Investments (FDI), and dynamics and growth of African business. Ebes lectured in Covenant University, Nigeria, before becoming a resident postdoctoral research fellow at UJ for 3 years. She won a best reviewer award from the international management division of the Academy of Management (AOM) in 2020. Before academia, she worked in the banking industry, and also holds a bachelor's degree with honors in accountancy from University of Nigeria, Nsukka, an MBA and MPhil from LBS-PAU, and a certificate in statistical analysis from the University of St. Gallen, Switzerland. Her research has been published, and presented on global platforms such as AOM, Strategic Management Society, and Business History conferences. In addition to research, Ebes founded and manages Student Founders and Entrepreneurial Platform (SFEP)—an organization that supports students in business and entrepreneurship.

A Strategic Human Capital Approach to the Sustainable Development of African Countries: Introduction

1

Abstract

This chapter serves as the introduction to the book *On the Sustainable Development of African Countries: A Strategic Human Capital Approach*. It provides the main purpose and specific objectives of the book, introduces the strategic human capital approach to sustainable development and its rationale, and explains why in comparison to other approaches, it is the superior path to sustainable development for African countries. This chapter also provides brief descriptions of the contents of the 3 parts and the other 11 chapters of the book.

Keywords

Sustainable development · SDGs · Human capital · Agenda 2063 · African countries

Africa is a continent that is home to 54 countries. Yet, Africa is often discussed as if it were one country, a homogeneous country with one culture. The diverse cultures, tribes, ethnicities, languages, and peoples of different races that make up the African continent are very often forgotten. Even countries within Sub-Saharan Africa (SSA), now construed as distinct from North Africa because the countries within the region appear to have similar characteristics, are very much diverse and different from one another. In fact, no two African countries can truly be regarded as the same. Each African country is unique almost in all respects, including in political and economic histories. Therefore, even when two countries were colonized by the same western country, each country's political history and the resultant political and socio-economic dynamics are truly really different. Still, one common factor that seems to bind all African countries together is developmental problems and challenges. From North Africa to Southern Africa, across West Africa through to Central and East Africa, African countries, especially those in SSA, are faced with similar developmental challenges.

Much has been written about the specifics of the developmental challenges of African countries which, though many and in myriad forms, all seem to be comparable across countries. Various global indexes that indicate economic growth and competitiveness almost always show African countries at the bottom; they are rarely at the top. Many reasons have been offered for Africa's developmental challenges. Among developmental economists, the reasons range from geographical, historical, cultural, and tribal factors to institutional factors (Moyo, 2009). The unfavorable geographical landscape; the peculiar colonial and other histories; the huge diversity in culture, tribe, ethnicity, and language that seem to hinder efforts at a holistic unity; and the lack of good governance and the institutions of quality governance have all been given as reasons by different economists for the inability of the continent to make good developmental progress (Moyo, 2009). These reasons align with those given by many others as the causes of poor economic growth and development of African countries.

Long-term economic growth and development depend on quality institutions and human capital (Acemoglu et al., 2014; Acemoglu & Robinson, 2013; Diebolt & Hippe, 2019; Newman et al., 2016; Rodrik, 1997); the primacy of one over the other is still arguable. However, one thing is clear that both sides of this particular argument are agreed on the importance of human capital for a country's economic growth and development. Several solutions have also been proffered on overcoming the continent's challenges, most of which somewhat correspond with the diverse reasons given for the developmental challenge in the first place. However, the solutions have somehow not yielded the expected results in solving the continent's developmental problems and challenges—of course, the assumption here is that the proffered solutions have, at the very least, been attempted.

Despite the ongoing development challenges of African countries, there are some positives on many fronts. Africa is home to some of the fastest growing global companies, and many of the banks on the continent have become global and can compete with their global rivals (Adeleye et al., 2018; Amankwah-Amoah et al., 2018; Verhoef, 2017). Kenya is the world leader in mobile money globally. Not too long ago, for a while, Africa seemed to have captured many global headlines with captions such as "the new frontier of growth" and similar captions. At this not-too-distant past, Africa was regarded as "the world's fastest-growing continent" (The Economist, 2013). Indeed, it was. Between 2001 and 2014, the gross domestic product (GDP) of African countries grew at an annual average of 5% (World Bank, 2021). However, much of this growth was from increase in prices of commodities and exports of primary products and natural resources. *The Economist* put the estimate of the

portion of the continent's GDP growth during this period that was from increase in prices of commodities as one-third (The Economist, 2013), which is roughly about 33%. Another estimate by the global management consulting firm, McKinsey, puts the growth from natural resources and the government spending they financed within the period at 32% (McKinsey, 2010), a figure close to the estimate from *The Economist*. Within this same period of economic growth, on the average, African countries had the highest export product concentration index globally.[1] This ranged from 0.311 in 2001 to 0.361 in 2014, reaching a high of 0.469 in 2008, compared to the world average of 0.068 in 2001 and 0.078 in 2014 (UNCTAD, 2024). The data for African countries is conspicuously higher than that of countries in other world regions.

Jointly, these data and statistics are a signal of the reliance of most African countries on the export of natural resources in their raw states and the consequent dependence on global commodities' prices. Although GDP growth on the continent has since waned, just as that of other global climes also have, the much-famed favorable headlines about Africa were not out of place. The primary commodity–induced economic growth is a pointer to the continent's abundant wealth in natural resources. Africa has 30% of the world's mineral reserves, 40% of the world's gold, and 90% of the world's chromium and platinum reserves (UNEP, 2023). In fact, Africa is host to some of the largest reserves of many of the minerals that are used to produce low-carbon renewable energies (UN, 2022a). Africa also has the world's largest arable landmass in addition to forests, woodland, freshwater resources, and wildlife (AFDB, 2016). Therefore, it is not surprising that reference is sometimes made to the continent as a resource-rich continent. However, the world and African countries themselves seem to take little notice of the continent's greatest wealth—

---

[1] This index measures the degree of concentration of goods, excluding services, exported by a country. The higher the index, the higher the export product concentration of each country signifying non-diversification in exports of goods.

the massive and growing youthful population, Africa's great potential for human capital.

Africa's Neglected Resource

According to available data from the World Bank (2022), the population of African countries was estimated to be over 1.4 billion[2] as of 2021. However, the interesting thing about the continent's population is not its current size. After all, Asia still has the highest population among the world's regions. The population of African region is second to that of Asia. The exciting thing about Africa's population is the age of the populace. Africa has the youngest population in the world. Data from the United Nations shows that about 70% of the people in SSA are under the age of 30 (Kamer, 2022; UN, 2022b). This is in sharp contrast to many countries, notably in the west, where a large proportion of the population are edging toward retirement age. With these population statistics, coupled with relatively high birth and fertility rates, compared to many other countries, African countries are poised to continue to have the youngest population in the world for many years to come. Africa has the highest number of countries with the highest population growth rates. The continent's population, all things being equal, is expected to double to about 2.5 billion people by 2050 (The Economist, 2020). Summarily, Africa not only has a huge population; it currently has the highest number of young people globally and will continue to do so in the near future. By the turn of the century in 2100, Africa is projected to have a population of 4 billion, two out of every five people on earth are expected to be Africans, and the median age in Africa is expected to be 35 years (Stanley, 2023; UN, 2022c). This means that by 2100, about 2 billion Africans are expected to be below

35 years. The year 2100 is only a few decades from now.

What do the above statistics portend for the continent? Is there potential for a demographic transition that could yield demographic dividends? Or is there more likelihood of a demographic doom? The answers to these questions are neither direct nor straightforward. Africa's greatest resource and potential human capital is its population and by extension its people who are potentially a great human capital resource. However, the continent's greatest resource has been neglected for too long. There seems to have been little significant and deliberate strategic development of the continent's population. Most positive discourses on Africa's youthful and growing population are almost always limited to its market size potential and the need to create jobs. While these are not out of place, a large population is not just about market size potential, and job creation will not happen accidentally. A large population is also, principally, about human capital potential. It is about the potential of people as resources to have diverse kinds of knowledge, skills, capabilities, and characteristics that enable them to live productive lives, provide creative solutions to problems, make significant contributions to the economy, and ultimately ensure sustainable development on all fronts. This other side of the coin of looking at a huge population size changes the perspective from merely the potential for consumption to possibilities for production and job creation. One of the ways to move these possibilities to realities is through human capital development. It is human capital, not natural or mineral resources, that foster creativity, productivity, competitiveness, and job creation. The reservoir of natural resources requires human capital to bring out the potential and inherent value that is often not apparent in natural resources. Without the requisite human capital–based developmental strategy, African countries will continue the export of primary goods and import same in the form of expensive and refined value-added products. Focused strategic human capital development is largely missing from the discourse on Africa's huge and

---

[2]Actual population was 1,470,564,145, made up of the figure for Sub-Saharan Africa and the six Northern African countries of Algeria, Egypt, Libya, Morocco, Sudan, and Tunisia, as data for Northern Africa as a block was unavailable.

growing youthful population both within and outside the continent.

For many African governments, the focus of many basic solutions to most of their developmental challenges seems to be that of reducing the plaguing infrastructural deficit. Consequently, roads and bridges continue to be built, and everything resembling physical infrastructure seems to be the emblem of economic growth and development. This perspective is not much different from that of the average African on the street. Suffice to say, many of the solutions almost always can be summarized into how to reduce the infrastructural deficit that plagues African countries. A focus on physical infrastructure is undoubtedly relevant and indeed ensures development in some aspects. As already mentioned above, it is well established in economics research, particularly in endogenous growth theory, that long term economic growth and development result from the fundamentals of quality institutions, physical capital of which physical infrastructure is part, and human capital, or what some refer simply to as human resources (Acemoglu et al., 2014; Acemoglu & Robinson, 2013; Newman et al., 2016; Rodrik, 1997). These fundamentals require time and consistent effort and commitment in investments for their impact to be realized. African countries have generally tended to emphasize the route of building physical infrastructure to the detriment of the other fundamentals of economic growth and development.

Physical infrastructural development in itself, and of itself alone, does not and cannot guarantee sustainable development. Electricity, transportation, telecommunications, and other infrastructure are definitely needed and indeed do contribute in the short and long term to economic growth and development (Sachs, 2005; Timilsina et al., 2023). However, physical infrastructure, regardless of how lofty such projects are and regardless of their novelty, beauty, and functionality, do not equate to sustainable development; on their own, they cannot guarantee any economic growth or development. Africa's development rests on ensuring that the physical infrastructure being built facilitates human capital development and deployment in one way or

the other. African countries need the combination of basic infrastructure and human capital development that can eventually culminate in sustainable economic development. The basic infrastructure that Africa needs are those that facilitate human capital development and human capital deployment such as those that enable energy supply and production and those that link up the countries and industries with global value chains. A focus on the development of physical infrastructure that does not facilitate human capital development or deployment is not going to lead to sustainable economic growth and long-term development. Physical infrastructure built merely for aesthetics will not serve Africa's sustainable and long-term development goals.

Moreover, all physical infrastructure are subject to diminishing returns on all fronts. On one hand, physical infrastructure requires continuous maintenance. They also require upgrade from time to time. On the other hand, it is people that make use of the built infrastructure. The lack of the requisite knowledge of the right use and maintenance of infrastructure ultimately increases maintenance costs and leads to frequent dilapidation. Optimal use and maintenance of physical infrastructure requires an educated populace, a populace with the requisite human capital. Consequently, sustainable development is not about building physical infrastructure, and this is aptly reflected in the United Nations Sustainable Development Goals (SDGs) which consists of a set of 17 broad and diverse goals.

The Strategic Human Capital Approach to Sustainable Development

Economic development is about human capital development and the institutions in place such as those that aid economic development which include those that enable quality governance and those that encourage investment in human capital development (Acemoglu et al., 2014; Goldin, 2016). Education and health care, the two main components of human capital, are necessities that enable productive economic activities (Delgado et al., 2012; Sachs, 2005). This book presents a strategic human capital approach to the sustainable development of African countries, the merits of such an approach, and details of what such an approach might look like. While the approach is generally anchored on the SDGs and African Union's Agenda 2063, it extends past these to the long-term economic growth and development of African countries beyond 2030 and 2063.

A strategic human capital approach to sustainable development emphasizes a focus on human capital—its accumulation, development, and deployment to economic productive activities. It is the deliberate and coordinated investment in the education and health of the people and the assessment and prioritization of other investments by their relations and impact on the productive capacities and well-being of the people. It is an approach that prioritizes investments based on their contribution to either human capital accumulation, development, or deployment. Under this approach, energy and physical infrastructure projects, for examples, become important because they facilitate human capital development and deployment by facilitating production and productivity and not merely for their esthetic value. In addition, given the peculiar case of African countries—the continent is almost the last frontier of economic development and the path dependencies that have resulted from the colonial past—and the overall trajectory of world events such as the emergence of digital technologies and digitalization, the strategic human capital approach is the feasible and plausible path to sustainable development for African countries. This book focuses on how African countries can strategically develop their human capital to unleash their countries' potential using this approach. Among other recommendations, the approach includes prioritizing investments in quality health care and education systems capable of delivering the knowledge, skills, and abilities for productivity in the modern economy of today and the future. Quality health care and conventional and digital literacy levels of many African countries are currently among the lowest globally. A strategic human capital approach incorporates policies and systems than can alleviate extant poor levels of quality health care and education. The approach is one that also aligns with sustainable economic and social development as outlined in the SDGs.

United Nations' SDGs and Africa Union's Agenda 2063

The SDGs The SDGs were adopted by the UN in 2015 and consist of a set of 17 goals in different broad developmental areas. They are a replacement for the eight Millennium Development Goals (MDGs) which were adopted at the turn of this century in the year 2000. Using 1990 as a base year, the MDGs aimed to reduce extreme poverty and child mortality and improve maternal health and combat HIV/AIDS, malaria, and other diseases, among other goals, all by the year 2015 (UN, 2015). The MDGs themselves were sequel to Agenda 21 which was adopted in 1992 with the broad aim of promoting global sustainable development. Hence, the collective global agenda toward sustainable development, a sort of development that takes cognizance of the effects of economic growth and production on the planet and future generations, started long before the SDGs were adopted in 2015. Overall, the SDGs aim to simultaneously end poverty and ensure that people enjoy peace and prosperity while protecting the planet (UNDP, 2024). An agenda to limit the impact of climate change is, therefore, an inherent part of the SDGs and sustainable development. Similar to the MDGs where each goal had underlying targets, within each SDG are a set of more specific goals also regarded as targets which are measured using certain indicators. Altogether, the 17 SDGs are associated with 169 targets and over 230 indicators. The SDGs build on the substantial foundational progress attained through the MDGs and previous global agendas toward sustainable development such as Agenda 21 (Table 1.1).

Despite these past global agendas, Africa's developmental challenges continue to subsist. A large portion of the over 700 million people still living in extreme poverty, that is, on less than $2.15 per day, for example, are in SSA (World

Table 1.1 Overview of the SDGs and the prior MDGs

	MDGs	SDGs
General aim and focus	End extreme poverty; global partnership between developed and developing countries	Peace, prosperity, and protection of the environment
Year adopted	2000	2015
Number of goals	8	17
Number of targets	18	169
Number of indicators	48	248; revised to 231 unique indicators
Broad goals	Goal 1: Eradicate extreme poverty and hunger Goal 2: Achieve universal primary education Goal 3: Promote gender equality and women empowerment Goal 4: Reduce child mortality Goal 5: Improve maternal health Goal 6: Combat HIV/AIDS, malaria, and other diseases Goal 7: Ensure environmental sustainability Goal 8: Develop a global partnership for development	Goal 1: No poverty Goal 2: Zero hunger Goal 3: Good health and well-being Goal 4: Quality education Goal 5: Gender equality Goal 6: Clean water and sanitation Goal 7: Affordable and clean energy Goal 8: Decent work and economic growth Goal 9: Industry, innovation, and infrastructure Goal 10: Reduced inequalities Goal 11: Sustainable cities and communities Goal 12: Responsible consumption and production Goal 13: Climate action Goal 14: Life below water Goal 15: Life on land Goal 16: Peace, justice, and strong institutions Goal 17: Partnerships for the goals
Proposed year of attainment of goals	2015 (past)	2030

Source: Author's compilation

Bank, 2024). With only a few years to 2030, the ability of African countries to achieve the SDGs by the given period target seems an almost insurmountable task. Africa's quest for sustainable development is sure to continue well beyond the end period of the current SDGs. Interestingly, before the SDGs were articulated in 2015, African Union's Agenda 2063 had already been put in place.

Agenda 2063 In 2013, the African Union (AU) put together a strategic framework which it states is aimed at repositioning the African continent to become a dominant global player by 2063. Termed Agenda 2063, and accompanied with the slogan, "The Africa We Want," its twin aims are inclusive and sustainable development. Agenda 2063 consists of 7 broad aspirations, 20 narrower

goals, and about 39 more specific goals in particular areas named as priority areas by the AU. Each broad aspiration has its own stated goal(s) that are generally narrower in scope than the aspirations, and each goal has its own still narrower priority areas meant to be the focus of each African country. Summarily, by the aspirations of Agenda 2063, by the year 2063, Africa desires to be a peaceful, prosperous, integrated, united, well-governed, people-driven, and globally influential continent. These aspirations are somewhat in alignment with some of the SDGs which are, arguably, a bit more specific than the rather broad but notable aspirations of Agenda 2063 because of the several measurable indicators included in the SDGs. Also included in Agenda 2063 are flagship projects and programs meant to propel the attainment of the stated aspirations, goals, and priority areas.[3] Clearly, Agenda 2063, which was adopted prior to the adoption of the SDGs, has a longer-term focus and development agenda as it extends well beyond 2015 to 2063. Agenda 2063's 50-year time frame is divided into a series of five 10-year implementation plans with the first period having elapsed in 2023. Although this strategic initiative of the AU is for African countries, the SDGs and Agenda 2063 both aim for sustainable and long-term development for African countries and, by extension, societies in Africa.

Why Does Human Capital Matter for Sustainable and Long-Term Development?

From history, Africa has not always been underdeveloped. In fact, history tells us that development started from Africa. Early development in Africa culminated and is evidenced in the several civilizations, kingdoms, and empires that characterized precolonial Africa. Many of these still remain across African countries, but modern development, as it currently is, seems to have eluded or is eluding African countries.

More than any other factor, modern economic growth and development is hinged on people: people's human capital, the knowledge, skills, abilities, and other characteristics of individuals that can be put to productive use. Human capital can enable innovation and entrepreneurship and holds so many benefits to individuals and the society. Human capital development directly relates to SDG 1, no poverty; SDG 2, zero hunger; SDG 3, good health and well-being; SDG 4, quality education; and SDG 8, decent work and economic growth, and indirectly relates to almost all the other SDGs. Investments in people invariably encompass providing quality education and investing in the health and well-being of people. In an increasingly knowledge and digitally driven economy, ensuring that people are able to get decent work certainly involves the requisite

development of human capital. The human capital approach is also amenable to the aspirations of Agenda 2063, one of which is to have an Africa that is people-driven. An Africa that is people-driven, as explicitly articulated in the sixth aspiration of the agenda, surely has to be one that puts people, and the development of their human capital, at the core of all sustainable and long-term development efforts.

The quality of human capital available is a prerequisite for the use to which other factors such as land, and by extension natural resources, can be put. It is people that determine the value of natural resources. It is also not very readily realized that technology, and all its advancements in its various forms and other forms of capital, are all products of the knowledge, skills, and abilities of people. The individual person is thus the most important resource of any country, far and above every other resource. Failure to develop the human person, as an individual and as a group of persons, and to harness the potentials of individuals for the individual's benefit, for the different organizational forms in the society and country, and for the society and country often results in different ails and ultimately in underdevelopment.

Structure and Contents of the Book

Through an interdisciplinary approach, this book presents the merits of a strategic human capital approach to sustainable development. Knowledge from multiple disciplinary perspectives are integrated to first enhance a holistic understanding of the concept of human capital and how it can be developed for the benefit of the individual owners and the organizational groups and societies to which they belong. The book then goes further to present practical theoretically based guidelines on how countries can develop and accumulate human capital for national development.

The 12 chapters of this book are grouped into three main parts. The first chapter, this chapter, serves as an introduction to the book. Part 1 lays the theoretical foundations of the concept of human capital in three different chapters. The

---

[3]Please refer to the AU for more details on the aspirations, goals, and priority areas.

second chapter in Part 1, provides the history, definitions, and meanings of human capital. The chapter traces the history of human capital from when the idea of a person as capital was largely frowned upon to the early conceptual and empirical studies of the concept, to when it became more acceptable. To provide a deeper understanding, the next chapter looks at the nature of human capital by examining the components, types, sources, and levels of human capital. The outcomes and benefits of human capital to the individual, and the various groupings to which the individual belongs, are presented in the fourth chapter and concludes the first part of the book. Chapters 2, 3, and 4 draw heavily from research and scholarship in economics, strategic human resource management, organizational behavior, business strategy (strategic management), and entrepreneurship. Attempt has been made to present these chapters in a manner that is easily readable and digestible to readers with nonacademic backgrounds. A key notion in this first part of the book is the recognition that just as individuals can have knowledge, skills, and abilities (human capital), groups of organizations of people such as firms, cities, societies, and countries can be embodied with group human capital. This notion that groups of people can have certain forms of human capital, as a group, is oftentimes not yet readily recognized although it is important for strategic planning. The notion is also important for the approach to sustainable development emphasized in this book.

There are four chapters in the second part of the book. The first chapter in Part 2, Chap. 5 of the book, focuses on some global human capital indexes that attempt to measure the human capital of countries. Chapter 5 also serves as a general introduction to the second part of the book which consists of case studies of some selected countries from three of the world's continental regions with great overall human capital formation and development: Asia, Europe, and North America. The last three chapters in this part of the book focus on short descriptive case studies of the human capital developmental paths of Singapore, Finland, and Canada. Each of the chapters focuses on a selected country from a continental world region. The short case studies are devel-

oped from publicly available materials and extant research. Countries were selected from each continental region based on their top placement and advancement, over the years, in the four main global human capital indexes: World Bank's Human Capital Index (HCI), the World Economic Forum (WEF)'s Global Human Capital Index (GHCI), United Nations' Human Development Index (HDI), and the Global Talent Competitiveness Index (GTCI) published by Insead Business School, France. Other global human capital indexes apart from these 4 are also discussed in Chap. 5.

For the case studies, selected countries were chosen for their consistent top rankings on the indexes and as leaders on the indexes in their continental region. Selected countries were also chosen to each showcase a particular area of human capital developmental path. The vignettes aimed to investigate the human capital strategies of these countries and to draw probable lessons from them. These mini-case studies highlight an important point that runs throughout this book— human capital development and the unleashing of its inherent potential has to be deliberately strategic—that is a coordinated approach to investing in the education and health of the population. Human capital accumulation and development and the utility from its deployment are not accidental. Realizing societal gains and benefits from human capital just don't happen. There has to be a plan, and that plan begins when countries have a vision of what they want to achieve. Human capital is then incorporated into the strategic vision and plan for its attainment. However, the mini case studies are not presented as examples to be replicated. The lessons from them are not met to be followed precept by precept by African countries or indeed any other country for that matter. They are presented as examples of outcomes of deliberate and strategic formation, development, and utilization of human capital.

Arguably, the most important part of the book is Part 3 as it delves into presenting the details of the strategic human capital approach and includes suggestions and recommendations for public policy for African countries. It is made up of four chapters. The first chapter in this final part of the book, Chap. 9, takes a look at the state of human capital in Africa and the potential given the current and projected demographics of African countries. Chapter 10 presents the basic generic strategic approaches to human capital accumulation that are available to countries. These approaches serve as a fundamental basis for the development of the policy recommendations in the next chapter. Chapter 11 provides specific policy directions on education, health, and the role different stakeholders can play in human capital development, among other recommendations and suggestions. This chapter is replete with policy agendas that can be adapted by African countries to strategically develop their human capital. The policy guidelines are embodied with in-depth explanatory notes that expound the rationale behind each general proposal. While Chap. 11 is targeted at policy makers, the policy frameworks are not limited to what only African policy makers can do. There are also suggestions and recommendations on how individuals and private and third sectors can be part of the human capital formation and development process. Finally, the aim is not for countries to simply adopt these recommendations and suggestions. No. The context of each African country is unique. One objective is for countries to adapt them to their peculiar circumstances and contexts or perhaps even glean some viable lessons from them. Another objective is for the agendas to spur the different stakeholders in African countries to begin to reflect and develop ways and means of investing directly and indirectly in the resources that matter most for the continent, the people.

The book concludes with Chap. 12. The chapter succinctly summarizes the contents and the different ideas presented and reiterated throughout the book. The three-part structure of the book, though suitable for all readers, is primarily directed toward specific readers. Researchers, academics, and students may be more adept at Part 1, due to its technical nature. Readers interested in global human capital indexes and the practical illustrations of how some countries have strategically developed their human capital may go directly to Part 2. Readers interested in human capital and its development in African countries will find Part 3 very enlightening. However, over-

all, readers interested in human capital will find great utility in all the parts of the book. Although each chapter builds on the one before it, each chapter is written as an independent, "standalone" chapter that can be understood on its own merit.

Concluding Remarks

In writing this book, the hope is that at the very least, African countries go beyond the focus on natural resources and the building of physical infrastructure for the mere fun of it and for just esthetic value and get to start thinking more deeply about their long-neglected resource: the people and their great potential as human capital resources. Island countries such as Singapore and Japan, with little or no natural resources, have managed to become developed and advanced countries by focusing on the development of their human resources. Even if the policy recommendations and suggestions presented are ignored, the other expectation is that the most important resource African countries hold, the people, will no longer be neglected. It is time for African countries to realize that effective governance is ultimately about people, and the human capital embodied in them, and not necessarily about the natural resources on and beneath the ground.

References

Acemoglu, D., & Robinson, J. A. (2013). *Why nations fail: The origins of power, prosperity, and poverty*. Crown Currency.

Acemoglu, D., Gallego, F. A., & Robinson, J. A. (2014). Institutions, human capital, and development. *Annual Review of Economics, 6*(1), 875–912.

Adeleye, I., Ngwu, F., Iheanachor, N., Esho, E., Oji, C., Onaji-Benson, T., & Ogbechie, C. (2018). Banking on Africa: Can emerging Pan-African banks outcompete their global rivals? In I. Adeleye & M. Esposito (Eds.), *Africa's competitiveness in the global economy* (pp. 113–136). Palgrave Macmillan.

AFDB. (2016). *African natural resources center: Catalyzing growth and development through effective natural resources management*. Available at https://www.afdb.org/fileadmin/uploads/afdb/Documents/Publications/anrc/AfDB_ANRC_BROCHURE_en.pdf

Amankwah-Amoah, J., Boso, N., & Debrah, Y. A. (2018). Africa rising in an emerging world: An international marketing perspective. *International Marketing Review, 35*(4), 550–559.

Delgado, M., Ketels, C., Porter, M. E., & Stern, S. (2012). *The determinants of national competitiveness.* NBER Working Paper Series 18249.

Diebolt, C., & Hippe, R. (2019). The long-run impact of human capital on innovation and economic development in the regions of Europe. *Applied Economics, 51*(5), 542–563. https://doi.org/10.1080/00036846.2018.1495820

Goldin, C. (2016). Human capital. In C. Diebolt & M. Haupert (Eds.), *Handbook of cliometrics.* Springer Verlag.

Kamer, L. (2022). *Median age in Africa 2022, by country.* Statista report.

McKinsey. (2010). *What's driving Africa's growth.* Available at https://www.mckinsey.com/featured-insights/middle-east-and-africa/driving-african-growth

Moyo, D. (2009). *Dead aid: Why aid is not working and how there is a better way for Africa.* Farrar, Straus and Giroux.

Newman, C., Page, J., Rand, J., Shimeles, A., Soderbom, M., & Tarp, F. (2016). Can Africa industrialize? In C. Newman, J. Page, J. Rand, A. Shimeles, M. Soderbom, & F. Tarp (Eds.), *Manufacturing transformation: Comparative studies of industrial development in Africa and emerging Asia* (pp. 257–276). Oxford Academic.

Rodrik, D. (1997). *Trade policy and economic performance in in Sub-Saharan Africa.* Expert Group on Development Issues, Harvard University.

Sachs, J. D. (2005). *The end of poverty: Economic possibilities of our time.* Penguin.

Stanley, A. (2023). *African Century.* IMF: International Monetary Fund. Available at https://policycommons.net/artifacts/4809842/african-century/5646416/ on 26 Jan 2024. CID: 20.500.12592/n4xg81.

The Economist. (2013). *Aspiring Africa.* Available at https://www.economist.com/leaders/2013/03/02/aspiring-africa

The Economist. (2020). *Africa's population will double by 2050.* Available at https://www.economist.com/special-report/2020/03/26/africas-population-will-double-by-2050

Timilsina, G., Stern, D. I., & Das, D. K. (2023). Physical infrastructure and economic growth. *Applied Economics.* https://doi.org/10.1080/00036846.2023.2184461

UN. (2015). *The Millenium development goals report.* Available at https://www.un.org/millennium-goals/2015_MDG_Report/pdf/MDG%202015%20rev%20(July%201).pdf

UN. (2022a). *Uncertain times, unsettled lives: Shaping our future in a transforming world.* Human Development Report 2021/2022.

UN. (2022b). *Young people's potential, the key to Africa's sustainable development.* Available at https://www.un.org/ohrlls/news/young-people%E2%80%99s-potential-key-africa%E2%80%99s-sustainable-development

UN. (2022c). *World population prospects.* Available at https://population.un.org/wpp/Download/Standard/Population/

UNCTAD. (2024). *UNCTAD Statistics: Empowering development through data as at 31st July, 2024.* Available at https://unctadstat.unctad.org/datacentre/dataviewer/US.ConcentDiversIndices

UNDP. (2024). *The SDGs in action.* Available at https://www.undp.org/sustainable-development-goals

UNEP. (2023). *Our work in Africa.* Available at https://www.unep.org/regions/africa/our-work-africa

Verhoef, G. (2017). *The history of business in Africa: Complex discontinuity to emerging markets.* Springer International.

World Bank. (2021). *GDP growth (ANNUAL %).* Available at https://data.worldbank.org/indicator/NY.GDP.MKTP.KD.ZG

World Bank. (2022). *Population, total.* Available at https://data.worldbank.org/indicator/SP.POP.TOTL

World Bank. (2024). *Understanding poverty.* Available at https://www.worldbank.org/en/topic/poverty/overview

Part I

Human Capital: Laying the Foundation

What Is Human Capital?

Abstract

The purpose of this chapter is to lay a foundation to understanding human capital. While there are several definitions of human capital, the definition of utmost utility to truly understanding human capital is the one that refers to human capital as the knowledge, skills, abilities, and other characteristics (KSAOs) of individuals that can be put to productive use and that have economic value. The concept of human capital inherently distinguishes human capital from labor also known as manual or commoditized labor. Consequently, all "labor" in the "labor" market are not the same. Human capital is that "labor" in the labor market that requires investment for it to be productive. Since economic value changes with time, context, and space, what can be regarded as human capital is therefore not static. Although the concept of human capital may have been "unthinkable" at a time in the distant past, the idea that humans have economic value is no more alien or offensive. It has become widely accepted, just like other concepts similar to human capital. This chapter explains the concept of human capital, its history, limitations, criticisms, and approaches to its study.

Keywords

Human capital · Commoditized labor · Knowledge · Skills · Abilities

Introduction

Before getting to the main focus of the book, it is expedient to understand the concept and the nature of human capital and to have a historical background of how the concept of capital relating to humans came to be. The term "human capital" is so ubiquitous these days that some organizations go as far as naming their human resource departments human capital department. Units and departments in organizations that were once commonly referred to simply as personnel departments have been replaced with names such as human capital department or strategic human capital department. However, even though the words "human capital" now appear so easily in various print, news, and online media; this has not always been the case. In fact, there was a time when it was almost a crime to combine and use the word "capital" with the word "human." Many decades ago, it was considered unethical and derogatory to refer to humans as capital. In other words, people could not be referred to as capital. So, what is human capital?

A cursory search for the meaning and definition of human capital unveils several definitions and meanings. This is partly because human capital is studied across many academic disciplines. Many scholars in economics, strategy, human resource management (HRM), organizational theory and behavior, international business, entrepreneurship, finance, psychology, market-

E. Aziegbe-Esho, *On the Sustainable Development of African Countries*, Sustainable Development Goals Series, https://doi.org/10.1007/978-3-031-81124-1_2

ing, and many other disciplines are students and scholars of human capital. However, the primary interests of scholars from various disciplines in human capital differ from one another. The interests of psychologists, for example, in the study of human capital, include the enhancement and development of psychometric tests as they attempt a better understanding of human beings and their cognitive abilities. Strategy scholars, on the other hand, study human capital as a form of firm resource that can be used to achieve the strategic vision or goals of the firm, to gain competitive advantage and improve performance. The several definitions of and meanings attributed to human capital also result partly from the ubiquity of the term in recent times. Nevertheless, the definition and meaning of human capital presented in this book is one which strives to cut across several academic disciplines. More importantly, it is one which, hopefully, can be easily understood by the average person: especially one that is neither a scholar nor affiliated to a particular discipline. First, before the definition of human capital, a brief look at the origins and history of the concept.

A Brief History of the Concept of Human Capital

Allusions to the concept of human capital can be found in the writings of many early scholars in economics dating as far back as William Petty in the seventeenth century AD (Kiker, 1966). Economists can therefore lay claim to originating the concept of human capital. Early economic scholars, such as Adam Smith, Jean Say, Friedrich List, Irving Fisher, and many others, vast not only in economics but also in philosophy and other scientific fields, all made allusions to the economic value of the skills acquired by human beings (Kiker, 1966). However, due to the derogatory and dehumanizing practice of slavery, many economists were divided on the notion and exact nature of human capital. While some regarded human beings as human capital, others regarded the skills of human beings as human capital. In the later notion, the person or the human is distin-

guished from the skills of the human (Kiker, 1966). This later notion recognizes, and seems to better preserve, the property rights and other human rights of the individual.

Adam Smith in his classic work, "An Inquiry into the Nature and Causes of the Wealth of Nations" originally published in 1776, saw the acquisition of talents through "education, study, or apprenticeship" as a form of (human) capital. In his written words,

> ...the acquisition of such talents, by the maintenance of the acquirer during his education, study, or apprenticeship, always costs a real expense, which is a capital fixed and realized, as it were, in his person. Those talents, as they make a part of his fortune, so do they likewise that of the society to which he belongs. The improved dexterity of a workman may be considered in the same light as a machine or instrument of trade which facilitates and abridges labour, and which, though it costs a certain expense, repays that expense with a profit. (Adam Smith, [1776] 2009: pg. 205)

It is clear from the above excerpt that Adam Smith recognized the distinction between the individual person and the skills acquired. He also recognized the difference between commoditized labor and human capital. Commoditized labor can aptly be described as the manual exertion of energy or power by humans in the execution of tasks and is unskilled. The assumption is that there is little or no cost to hiring, losing, and replacing employees with commoditized labor (Sevcenko & Ethiraj, 2018). Human capital, on the other hand, goes far beyond the manual exertion of power. It cannot be denied that all tasks require some form of manual exertion of power. However, human capital facilitates labor by reducing the manual exertion of power that is required to perform tasks. In this light, human capital can be seen as the combination of the "intangible force" that reduces the required amount of physical exertion that is necessary to perform tasks. Seen from this perspective, human capital requires some investments to either create or increase this "intangible force."

Also, as far back as 1890, Alfred Marshall made reference to the concept by calling it the capital that is invested in human beings (Goldin & Katz, 2024). It is not precisely clear on the first economist to coin the phrase "human capital." There are different accounts of when the phrase "human capital" was first coined and the two words used jointly. Irving Fisher is said to have made reference to human capital in his writings by using the words "the living capital" (Goldin, 2016) and even actually using the precise phrase "human capital" (Goldin & Katz, 2024). Arthur C. Pigou distinguished between investment in human and material capital in 1928 and is also credited to be the first to use the two words, human capital (White, 2017). In the immediate subsequent years following 1928, many prominent scholars in economics such as Roy Harrod, Frank Knight, Kenneth Boulding, Milton Friedman, and Joseph Spengler made significant mentions of the term in some of their writings (Teixeira, 2014). Mincer's (1958) article titled "Investment in Human Capital and Personal Income Distribution" also specifically mentioned the term "human capital." Mincer's article's main thesis was focused on analyzing the income differences that result from investments in human capital. However, not much was said about the nature of such investments, the nature of human capital itself, or how human capital creates value. Nevertheless, his contribution to the development of the emergence of the concept has resulted in several contemporary studies attempting to investigate the many types of returns to various forms of formal education and on-the-job training.

The nature of the investments in people and why such investments qualify as capital was further elaborated by Theodore Schultz in 1961 to reflect the increased capacity of human beings to earn and do more work beyond the physical exertion of energy. Theodore Schultz classified all the useful skills and knowledge acquired by individuals throughout their lifetime as human capital and indeed compared it to nonhuman capital when he wrote the following:

> Although it is obvious that people acquire useful skills and knowledge, it is not obvious that these skills and knowledge are a form of capital, that this capital is in substantial part a product of deliberate investment, that it has grown in Western societies at a much faster rate than conventional (nonhuman) capital, and that its growth may well be the most distinctive feature of the economic system…. (Schultz, 1961: pg. 1)

At the time Schultz's article was published in 1961, the concept of "humans" as "capital" or even as economic assets was neither popular nor readily acceptable. The term was so controversial that Schultz sometimes used the term "human wealth" instead of "human capital" in some of his earliest writings (Teixeira, 2014). Gary Becker was also rather hesitant to use the term in his earliest works on the subject. This was because referring to humans as capital reminded many of slavery where some men were treated as property rather than as men of free will. Before this time, research that explored the notion of human capital were scarce and far between. Scholars such as Walsh (1935), for example, and a few others such as Mincer (1958) explored the returns to investments in people primarily in the form of formal education such as schooling. Walsh (1935) compared the estimated costs of formal education to wages earned in a person's lifetime and found that wages earned in a lifetime generally exceeded the costs of formal education. He concluded that the concept of capital could be applied to man as the returns to formal education could be compared to the returns to other forms of capital. Similarly, Mincer (1974) also analyzed the impact of formal education and work experience on earnings and found that differences in earnings were largely accounted for by years of formal education and work experience. The special issue on human capital titled "Investment in Human Beings" by the *Journal of Political Economy* in 1962 served almost as a formal recognition of the concept of human capital. Therefore, the general idea of human capital stems from the notion that there is a return to investments in humans. There is a return to investments in people. Investments in people can truly be considered a form of capital "because it is the source of future earnings, or of future satisfactions, or of both of them. It is human because it is an integral part of man" (Schultz, 1972: pg. 5). It can also be the source of several societal benefits. Who should make this investment in people is still subject to debate. However, it has been empirically proven without any reasonable doubt that investments in humans is a form of capital capable of yielding benefits to the individual owners and to the group to which the individual belongs.

Nobel laureate Gary Becker's works on human capital in 1962 and 1964 provide the bedrock of human capital and much of classic human capital theory. Most studies on human capital in economics and other academic fields have been influenced fundamentally by Becker's theoretical analysis. Recent studies on human capital still make reference to Becker's works on the subject. Becker (1962), in one of the publications in the special issue on human capital by the *Journal of Political Economy*, presented a foremost, formal, and detailed theoretical analysis of human capital, some theoretical foundations, and some empirical implications. Summarily, Becker (1962) argued that investments in people in the form of education, knowledge, and health can be considered a form of capital as they increased individuals' incomes. Prior to this, the emphasis was on physical capital as the main source of individuals' incomes. Becker (1962, 1964) also distinguished between firm-specific human capital and general human capital. He referred to firm-specific human capital as human capital that can applied to productive tasks only in one specific firm, usually in the firm in which it was developed. Consequently, he referred to firm-specific human capital as human capital that results from training and experience in a specific firm. Firm-specific human capital is not easily deployable in other firms without corresponding loss in value. General human capital, on the other hand, is human capital that has equal value in productive tasks across all firms. In Becker's view, general human capital is widely deployable in other firms outside the focal firm in which it was developed and accumulated. Becker theorized that firm-specific human capital increases marginal productivity in the firm in which it is developed and the increase in productivity is shared between the firm and workers in the form of higher profits and wages, respectively. This higher wage would motivate workers to stay on at the focal firm in which they developed their firm-specific human capital. General, or generic, human capital, on the other hand, is equally useful across all firms and not capable of hindering workers' mobility to other firms.

Becker's classification of human capital into general and firm-specific was based on the specificity of human capital: a reflection of the application of human capital to productivity within a given firm context. Specificity of human capital is the reverse of the transferability of human capital to productivity across firms. The focus of Becker's theoretical analysis was on the forms of investment, such as education and on-the-job training, which he theorized were a form of intangible resource embedded in people with a potential to influence earnings, wages, and the productivity of firms.

The distinction between firm-specific and general human capital was used by Becker to examine the division of the costs and expected returns to investments in human capital between individuals and firms. However, this initial distinction made by Becker (1962, 1964) was mainly on how different kinds of investments in human capital would affect returns to such investments. Becker's concern about returns to general human capital and firm-specific human capital was used, among other issues, to determine when firms or individuals would pay for investments in general and firm-specific training. He argued that firms would be unwilling to pay for investments in general human capital because it is equally useful across many firms as the marginal increase in its productivity would be the same across all firms. Firms would however be willing to pay for investments in firm-specific human capital because it is most useful in the focal firm in which it is developed and would lead to increase in marginal productivity at the focal firm. The increased marginal productivity from firm-specific human capital would thus be shared between the firm and workers with such human capital. Workers would then be willing to stay at the focal firm because of increased wages which they would not be able to receive from other firms. An understanding of this rather brief summary of Becker's (1962, 1964) theoretical analysis is important not only for differentiating general human capital from firm-specific human capital but also for understanding other potential implications of human capital. There is more on this in Part 3 of this book, but this brief summary is essential to lay a historical foundation on human capital and its related theoretical analysis.

The main motive of early scholars of human capital was to establish the concept of human capital by answering some related questions. Were the skills acquired by humans through formal education a form of capital? Was it really beneficial for humans to acquire formal education? Was there a return to the costs of acquiring knowledge and skills through formal schooling? In other words, like all other forms of capital, could formal schooling and education be regarded as a form of investment to which returns could be expectedly made? Attempts to answer these questions led to both theoretical and empirical studies of human capital, although such studies were initially concentrated in the economics field and conducted by economists. The theoretical and empirical foundations for human capital were therefore laid in the 1960s (Sweetland, 1996).

Further Studies of Human Capital

Beyond the early research on the concept, later research work on causes of economic growth by Robert Solow and other economics demonstrate that a huge portion of modern economic growth could not be explained by physical capital and labor alone (Goldin, 2016; Solow, 1956, 1957). The impact of formal education significantly reduces the huge residual that could not be accounted for by the accumulation of physical capital and labor (Goldin, 2016; Mankiw et al., 1992). Human capital therefore contributes a significant portion of modern economic growth and is implicit in modern growth theories (Galor & Weil, 2000; Goldin, 2016). Although research on endogenous economic growth did not set out to prove the existence of human capital, it has nonetheless somewhat contributed to demonstrating the existence and impact of human capital.

Over recent years, human capital has become an accepted concept that is studied in many aca-

demic disciplines beyond its originating field of economics. Human capital forms the basis of much theoretical exploration and empirical research in many academic disciplines such as industrial and applied psychology, sociology, organizational behavior, strategic human resource management (SHRM), entrepreneurship, education, and strategic management. However, the motives for scholarship in human capital have evolved over the years and depend primarily on the main interests and objectives of the academic field of study. Although human capital is studied in different disciplines, interest in the concept has been from different perspectives. The interest of psychologists in human capital, for example, has traditionally been in the effective selection and development of human resources. Consequently, psychometric tests have been developed by psychologists for the assessment of human capital in an attempt to aid effective employee selection and development (DeNisi et al., 2003; Wright & McMahan, 2011). In the strategic management field, one of its main interests and objectives is the explication of differential firm performance (Rumelt et al., 1994). Therefore, the propelling overarching question that drives the interest of strategic management scholars in human capital is how human capital, as a resource to the firm, can be a source of competitive advantage and superior performance. Relatedly, human resource management scholars are interested in how employees with human capital can perform optimally while working in a firm. Indeed, there was a time units and departments in organizations responsible for staff were called as personnel units and departments. In more recent times, the names for such departments have changed and are now mostly referred to either as human resource departments or human capital departments. This evolvement in names from personnel department to human capital department shows the level of acceptance that the term, human capital, connotes even in practice and industry, not just within academic and scholarly circles. The concept of human capital has become so ubiquitous and accepted in industry and practice that some global agencies have put indexes in place to measure the human capital levels of countries:

World Bank's Human Capital Index (HCI), the World Economic Forum (WEF)'s Global Human Capital Index (GHCI), and United Nations' Human Development Index (HDI). Meanwhile, economists have continued to expand the study of human capital to investigating its effects on wages and wage differentials; employee mobility across occupations, industries, and countries; and other contemporary issues.

Summarily, the concept of human capital has become widely accepted. In fact, the concept has become so widely accepted that its meaning is sometimes lost. Many use the words human capital often times without cognizance of its actual meaning.

Definition(s) of Human Capital

Early scholars of human capital did not expressly define human capital although Schultz and Becker each later gave their definitions of human capital in their later works on the subject. Academic and other literature are however flooded with several definitions of human capital. Interestingly, several of the definitions of human capital currently available are indeed very similar. Research in management and organizational disciplines usually define human capital either at the individual or group level. At the individual level, it is defined as the knowledge, skills, abilities, and other characteristics (KSAOs) of individuals (Coff & Kryscynski, 2011; Crook et al., 2011). These other characteristics could be personality traits, affect, or behavior (Wright et al., 2014). Others define it at a group level, as a unit- or firm-level resource that results from the aggregation of the KSAOs of individuals in a unit or firm. In 2001, the Organization for Economic Co-operation and Development (OECD) defined human capital as the "knowledge, skills, competencies and attributes embodied in individuals that facilitate the creation of personal, social and economic well-being." OECD's definition touches on the broad categories of persons that can benefit from human capital. The World Economic Forum defined human capital as "the skills and capacities that reside in people and that

are put to productive use" (WEF, 2015). The World Bank adopts a more comprehensive definition as "the knowledge, skills, and health that people invest in and accumulate throughout their lives, enabling them to realize their potential as productive members of society" (World Bank, 2019). Table 2.1 presents some explicit definitions of human capital across some disciplines.

A summary definition of human capital, adopted for this purpose of this book, is that "human capital is the Knowledge, Skills, Abilities, and Other characteristics (KSAOs) of individuals that can be put to productive use and have economic value." The inclusion of other characteristics in this definition implies that there are several attributes of individuals that could be regarded as human capital. As reflected in the definitions in Table 2.1, cognition, education, experience, personality traits, affect, health, natural and learnt talent, know-how, aptitude, attitude, expertise, and behavior can all be human capital. While some of the characteristics of individuals that can be human capital, such as knowledge, cognition, experience, and education, can be regarded as relatively stable characteristics, others such as affect and behavior are more malleable. Stable characteristics are more difficult to change than malleable characteristics. This underscores why most definitions refer mainly to more stable characteristics. KSAOs may also be natural, physical, or innate. KSAOs are individually and cumulatively human capital resources. However, in practice, the acquisition and increase in any KSAO becomes merged with the pre-existing ones such that the resulting cumulative human capital becomes almost indistinguishable from its constituent past when put to productive use.

The summary definition in the preceding paragraph and the various definitions available in the literature imply that human capital can be at different levels. Indeed, this is at the heart of the difference in the various definitions. The various definitions also account for the different conceptualizations and associated variables of human capital in empirical studies. The wide range of variables that have been used to measure human capital includes formal education and schooling,

Table 2.1 Sample of definitions of human capital (primarily in economics and management)

No.	Publication/study	Academic domain of study	Level of definition	Definition of human capital
1.	Goode (1959: pg. 147)	Economics	Individual	"knowledge, skills, attitudes, aptitudes, and other acquired traits that contribute to production"
2.	Rosen (1989: pg. 136)	Economics	Individual	"the productive capacities of human beings as income producing agents in the economy"
3.	Schultz (1993)	Economics	Individual	"…the acquired abilities of people—their education, work experience, skills and health…"
4.	Huselid et al. (1997: pg. 171)	SHRM	Group	"…employees collective knowledge, skills, and abilities"
5.	Bontis et al. (1999: pg. 393; 397)	SHRM	Group	"the human factor in the organisation; the combined intelligence, skills and expertise that gives the organisation its distinctive character" "the collection of intangible resources that are embedded in the members of the organization"
6.	Hitt et al. (2001: pg. 14)	Strategy	Individual	"… human capital attributes (including education, experience, and skills)"
7.	Sharpe (2001: page 3)	Economics	Individual/ group	"the aggregation of investments in such areas as education, health, on-the-job-training, and migration that enhance an individual's productivity in the labour market, and also in non-market activities"
7.	Becker (2002: pg. 3)	Economics	Individual	"Human capital refers to the knowledge, information, ideas, skills, and health of individuals"
8.	DeNisi et al. (2003:pg 6)	Strategy	Individual	"…refers to all of the resources that individuals directly contribute to an organization"
9.	Hatch and Dyer (2004: pg. 1158)	Strategy	Individual	"Human capital begins with resources in the form of knowledge and skills embodied in people."
10.	Skaggs and Youndt (2004: pg. 86)	SHRM	Individual	"…high levels of skill, knowledge, and expertise"
11.	Youndt and Snell (2004: pg. 338–339)	SHRM	Individual	"The knowledge skills and experience of employees" "Individual employee's knowledge, skills, and expertise"
12.	Combs et al. (2006: pg. 502)	SHRM	Individual	"…employees knowledge, skills, and abilities (KSAs)"
13.	Somaya et al. (2008: pg. 936)	SHRM/ strategy	Individual	"the cumulative knowledge, skills, talent, and know-how of the firm's employees"
14.	Coff and Kryscynski (2011: pg. 1430)	Strategy	Individual	".. -an individual's stock of knowledge, skills, and abilities (hereafter called skills) that can be increased through mechanisms like education, training, and experience"
15.	Ployhart and Moliterno (2011: pg. 128)	SHRM/ strategy	Group	"… a unit-level resource that is created from the emergence of individuals' knowledge, skills, abilities, and other characteristics (KSAOs)"
16	Wright and McMahan (2011: pg. 95)	SHRM	Group	"…the aggregate accumulation of individual human capital that can be combined in a way that creates value for the unit"
17.	Jiang et al. (2012: pg. 1264)	SHRM	Individual	"…employees' competencies—that is, their knowledge, skills, and abilities"

(continued)

Table 2.1 (continued)

No.	Publication/study	Academic domain of study	Level of definition	Definition of human capital
18.	Campbell et al. (2014: pg. 533)	Strategy	Individual	"… knowledge, skills and abilities of individuals…."
19.	Crocker and Eckardt (2014: pg. 511)	Strategy/ SHRM	Individual	"Knowledge and skills that cannot be separated from individuals"
20.	Ployhart et al. (2014: pg. 374)	SHRM/ strategy	Individual and group	"Human capital resources are individual or unit-level capacities based on individual KSAOs that are accessible for unit-relevant purposes."
21.	Goldin (2016: pg. 22)	Economics	Group	"Human capital is the stock of productive skills, talents, health and expertise of the labor force,…"
22.	Bowlus et al. (2022: pg. 1)	Economics	Individual	"Human capital refers to any skills, competencies and knowledge workers have that potentially increase their productivity"
23.	Cappelli et al. (2023)	Economics	Individual	"the set of knowledge and skills that individuals accumulate over time"

Source: Author's compilation

training, different shades of experience such as employment, start-up, and owner experience, skills, and knowledge (Crook et al., 2011; Unger et al., 2011). However, human capital is not just people's KSAOs. The KSAOs have to have the capacity to be put to some productive use for the individual and the organizational context to which the individual belongs. Organizational context could be a business, society, country, or even a smaller unit within a formal organization or business.

Defining human capital in relation to KSAOs of individuals that can be put to productive use and have economic value brings to light the importance of organizational context. This is because economic value depends on time, context, and space. Indeed, because economic value depends on these three factors: time, context, and space, human capital is also dependent on time, context, and space. Therefore, the KSAOs of individuals that have economic value in one context, and that can be regarded as human capital, may not have economic value in another context. This somewhat relates to the specificity, and inversely the transferability, of human capital as earlier discussed. Certain forms of human capital may be immensely valuable in one organizational context, and at a certain time, and be of less economic value in another context and time.

Unfortunately, defining human capital in relation to economic value also obfuscates the meaning of human capital in another way—it confounds human capital itself with its effects, applications, and functions. This is because of the intangibility and invisibility of human capital. Although the economic value created from the application of human capital may be seen, the human capital responsible for creating the economic value cannot be seen physically. Only the individual or individuals with human capital can be seen. It can only be observed, and this observation is always in reference to the economic value that can be generated or the productive activity that it can be used for.

Perspectives and Approaches to Human Capital

As can be seen from the discussion of the history and origins of the concept of human capital, the concept and theory of human capital clearly originates in economics. However, human capital has also been studied and indeed can be studied using different lenses. This depends primarily on the field of study. However, in management field, specifically strategic management, to understand how value can be created from human, it is extremely useful to consider human capital as a resource. Human capital can be a resource to the individual embodied with it and to the collective group to which the individual belongs. Therefore, human capital can be a resource to a firm or an organization, a society, and a country. Human capital can create value, and therefore be a resource, to the individual owner and to the collective community and society to which the individual belongs. The knowledge, skills, abilities, and other characteristics of individuals with economic value can be a resource to an individual or a collective. The configuration and combination of KSAOs of individuals create value for them as individual persons or for the collective to which they belong. The recent approach of studying human capital resources (Ployhart et al., 2014; Ray et al., 2023), rather than just human capital, recognizes and emphasizes the collective resource that emerges from KSAOs of individuals at the individual, firm, and more collective levels. In their seminal work on this approach, Ployhart et al. (2014) distinguish human capital, "an individual's KSAOs that are relevant for achieving economic outcomes," from human capital resources, "human capital that are accessible for a unit's purposes" (page. 376). This approach, though somewhat conceptually different from the traditional concept, also emphasizes the emergence process that takes place within the context in which human capital is deployed to economic activities.

Health as Human Capital

Some of the definitions of human capital above explicitly include health, and it is implicit in those that do not expressly include it. Without good health, an individual and indeed a collective unit of individuals such as countries lose economic productive ability. Good health, both physical and mental, also allows people to acquire more education, skills, and other forms of human capital (Ridhwan et al., 2022). Consequently, the importance of health as human capital cannot be overemphasized. The significance of health as a component of human capital is reflected in sev-

eral extant studies that relate it to the economic growth of countries. In their recent meta-analysis of some extant studies, Ridhwan et al. (2022) find that health generally increases economic growth. Usually measured as life expectancy, the positive relationship between health and economic growth is however not linear (Acemoglu & Johnson, 2007; Bloom et al., 2019; Ridhwan et al., 2022). Neither is it straightforward and necessarily the same for developed and developing countries (Bloom et al., 2019; Ridhwan et al., 2022). Different dimensions of health also have different effects on economic growth, and the effect depends on a number of other factors. Nevertheless, the economic growth prospects of increased good health for African countries is great because of the potential demographic transition and demographic dividends that can result from the huge and growing youthful population of the continent.

Concepts Similar to Human Capital

Since the broad acceptance of human capital, other similar concepts have emerged. Concepts such as intellectual capital, structural capital, relational capital, and social capital have also become quite popular topics both in academia and in practice. Apart from these concepts being intangible assets, and therefore majorly invisible, they also share some commonalities that makes them forms of capital—investments in them can lead to production of value in some forms, and investments in them often yield a return, like all other forms of capital. They can also be accumulated over time and may also depreciate in value. However, unlike physical and tangible forms of capital, they cannot be purchased easily from factor markets. Hence, they have to be built or accumulated over time. As shown above in Table 2.1, these intangible forms of capital can be at the individual or group or collective level. Consequently, they can be accumulated by individuals, organizations, and even by communities and countries, although focus in the literatures is usually at the level of interest of the field of study.

In recent times, there has been the tendency of some to refer to "human capacity" rather than human capital. Adapting Danquah et al.'s (2022) definition of capacity, human capacity broadly refers to all the human capabilities that are necessary to assume responsibility for the improvement of an individual's life and environment. The focus of human capacity is beyond economic productivity and also extends to societal levels. In this sense, human capacity can be said to be broader than human capital. Conceptually, human capacity has also developed quite independently from that of human capital although much of the literature on it borrow from the definitions and frameworks of human capital. The use of capacity instead of capital also tends to avoid some of the limitations of human capital and the criticisms associated with referring to humans as capital.

Limitations of Human Capital

Unlike other forms of capital such as physical and financial capital like money, cash, and machines, human capital has one great limitation. It resides in people and is inseparable from the person that possesses it. Therefore, people can quit, ask for more compensation from the firm in which they deploy their human capital, or even lose the motivation to apply their human capital attributes to productive work (Coff, 1997). This limitation applies not only to workers in an organization but also to individuals themselves. People can simply refuse to work for themselves or for others in an organization even when they have the most sought after KSAOs. This limitation is intricately linked to the freewill of individuals who have the choice on what KSAOs they want to acquire or develop, how they want to apply and deploy the KSAOs, or whether they will even apply them at all. In the broadest context of a country, people have the freewill to choose the country in which they want to deploy their human capital to productive activity regardless of where they might have acquired or developed their human capital. Relatedly, individuals with very valuable human capital may also pos-

sess some other characteristics that are not so valuable (Ployhart et al., 2014). Some of these characteristics may even be undesirable. Human capital resides in people and come in the bundle in which the person comes in. Each individual is a portfolio of different KSAOs and characteristics. In addition to the human capital KSAOs that are desirable and valuable, there may be other KSAOs that are currently not valuable and desirable. This major limitation has several other implications including who should bear the costs and obtain the benefits of human capital investments.

Criticisms of Human Capital

Since the formal conceptualization of human capital, the concept, along with its theory, has faced many criticisms. As already mentioned earlier, there was reluctance to apply the concept of capital to humans because it had the connotation of slavery. Indeed, Schultz and Becker, two of the foremost pioneer initiators of the concept, admit to initial reluctance to use the term "human capital" (Becker, 1994; Breit & Hirsh, 2004; Tan, 2014; Teixeira, 2014). They contemplated using human wealth and human investment rather than human capital. Human capital was so bitterly criticized in the early days of its development that some critics referred to the term as "human cattle" (Tan, 2014), a reflection of a seeming negative connotation of the term.

Critics of human capital theory come from diverse disciplines. They argue that the theory fails the test of realism, has methodological issues, lacks empirical evidence, and is overly simplistic (Tan, 2014). Without delving deep into each criticism, this final section of this chapter presents a cursory summary of extant criticisms of the notion of human capital itself rather than its theory.

Tan (2014) presents a good summary of extant criticisms of human capital as falling into four main categories: methodological individualism, empirical, practical, and moral. The idea of human capital is seen as some as dehumanizing as it reduces humans to machines who are purely driven by utility and whose sole purpose is economic production (Drobny, 2017; Hyslop-Margison & Sears, 2006). Consequently, the term "human capital" gives the impression that humans are seen as being without souls or spirits, an issue that Becker himself had grappled with in coming up with the term (Breit & Hirsh, 2004; Tan, 2014; Teixeira, 2014). Humans are thus seen as being reduced to the level of capital, objects like other nonhuman capital (Drobny, 2017). Others argue that human capital places too much emphasis on the individual members of society—methodological individualism. Some also argue that it is difficult to actually measure human capital, and current measurements of the concept are overly simplistic (Allais, 2012; Winterton & Cafferkey, 2019). Some critics are, therefore, not necessarily against the notion of human capital. Rather, the argument is against most common extant measures of human capital that limit its measurement to only education and training.

Scholars from the field of education have been particularly skeptical of human capital, and human capital theory, because the notion of human capital reduces education to a mere business activity that is driven solely by profit rather than for other noble ideals such as self-enrichment, human development, and for proper functioning of "thick" democracy that facilitates civil liberties and effective governance (Ball, 2010; Marginson, 1997 in Tan, 2014; Hyslop-Margison & Sears, 2006; Nussbaum, 2010). In other words, human capital reduces education to a means to earning a living rather than for living (Hyslop-Margison & Sears, 2006). Consequently, it seems emphasis is placed on ensuring education meets the needs of the economy (Allais, 2012) and not seen as a fundamental human right that everyone should have (Edwards, 2018). Based on this, the argument is that human capital negates the fundamental human right to education that everyone should have. Other disciplines have also accused economists of overstepping their discipline's boundary and intruding into realms traditionally within the boundaries of other disciplines through human capital (Tan, 2014).

However, some of these arguments are not exactly contradictory or irreconcilable with human capital. Education, for example, as a fundamental human right and one which all governments should provide as a public good does not contradict the notion that it is simultaneously an investment in people which can yield private and public returns. Neither do governments' provision of education preclude provision by private members of the society—both are not mutually exclusive. The fact remains that education and training are investments in humans capable of yielding private and social returns to individuals and the society to which they belong. Now, the determinants of these returns; the manner in which they come, and if they come at all; and other related issues may be affected by other factors apart from "these" investments in people. However, this does not invalidate the notion that investments in people have value to the people and to the organizational contexts (firms, societies, and countries) to which they belong. Perceiving education as human capital capable of yielding economic returns does not preclude the other benefits that come with education. Consequently, applying an investment rationale to investments in people is only logical despite its limitations and possible negative connotations. Moreover, formal education is not the sole component of human capital, albeit a major and important source of it. In regard to the theory of human capital, like any other theory, it is a tool that helps in simplifying complex phenomena and relies on assumptions which may sometimes appear overly simplistic. However, the simplicity of the theory and the assumptions do not negate its utility for gaining insights into complex phenomena whose comprehension and understanding may have been impossible or overly difficult otherwise.

Conclusion

Human capital consists of the knowledge, skills, abilities, and other characteristics (KSAOs) of individuals that can be put to productive use and have economic value. Despite initial, and more recent, criticisms of the concept that investments in humans do indeed have a value akin to other forms of capital that are more tangible, the concept of human capital has become accepted. Indeed, it has become quite ubiquitous. Understanding the meaning of human capital is necessary to understanding its benefits and its role in the modern economy and society. It also lays the foundation to why a strategic human capital approach is the most feasible and best path for the sustainable development of African countries.

References

Acemoglu, D., & Johnson, S. (2007). Disease and development: The effect of life expectancy on economic growth. *Journal of Political Economy, 115*(6), 925–985. https://doi.org/10.1086/529000

Allais, S. (2012). 'Economics imperialism', education policy and educational theory. *Journal of Education Policy, 27*(2), 253–274. https://doi.org/10.1080/02680939.2011.602428

Ball, S. (2010). New voices, new knowledges and the new politics of education research: The gathering of a perfect storm? *European Educational Research Journal, 9*(2), 124–137.

Becker, G. S. (1962). Investment in human capital: A theoretical analysis. *The Journal of Political Economy, 70*(5), 9–49.

Becker, G. S. (1964). *Human capital: A theoretical and empirical analysis, with special reference to education*. University of Chicago Press.

Becker, G. S. (1994). *Human capital: A theoretical and empirical analysis, with special reference to education*. University of Chicago Press.

Becker, G. S. (2002). The age of human capital. In E. P. Lazear (Ed.), *Education in the twenty-first century* (pp. 3–8). Hoover Institution Press.

Bloom, D., Kuhn, M., & Prettner, K. (2019). *Health and economic growth*. Oxford Research Encyclopedia of Economics and Finance. Retrieved November 5, 2024, from https://oxfordre.com/economics/view/10.1093/acrefore/9780190625979.001.0001/acrefore-9780190625979-e-36

Bontis, N., Dragonetti, N. C., Jacobsen, K., & Roos, G. (1999). The knowledge toolbox: A review of the tools available to measure and manage intangible resources. *European Management Journal, 17*(4), 391–402.

Bowlus, A., Park, Y., & Robinson, C. (2022). *Contribution of human capital accumulation to Canadian economic growth*. Staff Discussion Paper, Bank of Canada.

Breit, W., & Hirsh, B. T. (2004). *Lives of the laureates: Eighteen Nobel economists*. MIT Press.

Campbell, B. A., Saxton, B. M., & Banerjee, P. M. (2014). Resetting the shot clock: The effect of comobility on human capital. *Journal of Management, 40*(2), 531–556.

Cappelli, G., Ridolfi, L., & Vasta, M. (2023). Human capital in a historical perspective. In *Oxford research encyclopedias*. Oxford University Press. https://doi.org/10.1093/acrefore/9780190625979.013.520

Coff, R. W. (1997). Human assets and management dilemmas: Coping with harzards on the road to resource-based theory. *Academy of Management Review, 22*, 374–402.

Coff, R. W., & Kryscynski, D. G. (2011). Drilling for micro-foundations in human capital-based competitive advantages. *Journal of Management, 37*, 1429–1443.

Combs, J., Liu, Y., Hall, A., & Ketchen, D., Jr. (2006). How much do high-performance work practices matter? A meta-analysis of their effects on organizational performance. *Personnel Psychology, 59*, 501–528.

Crocker, A., & Eckardt, R. (2014). A multi-level investigation of individual- and unit-level human capital complementarities. *Journal of Management, 40*(2), 509–530.

Crook, T. R., Todd, S. Y., Combs, J. G., Woehr, D. J., & Ketchen, D. J., Jr. (2011). Does human capital matter? A meta-analysis of the relationship between human capital and firm performance. *Journal of Applied Psychology, 96*(3), 443–456.

Danquah, J. K., Crocco, O. S., Mahmud, Q. M., Rehan, M., & Rizvi, L. J. (2022). Connecting concepts: Bridging the gap between capacity development and human resource development. *Human Resource Development International, 26*(3), 246–263. https://doi.org/10.1080/13678868.2022.2108992

DeNisi, A., Hitt, M. A., & Jackson, S. E. (2003). The knowledge-based approach to sustainable competitive advantage. In A. DeNisi, M. A. Hitt, & S. E. Jackson (Eds.), *Managing knowledge for sustained competitive advantage: Designing strategies for effective human resource management*. Jossey-Bass.

Drobny, P. (2017). The human as capital? A contribution to the critique of the theory of human capital. *Annales. Ethics in Economic Life, 20*(5), 95–106. https://doi.org/10.18778/1899-2226.20.5.08

Edwards, D. (2018). *What's wrong with the World Bank's human capital index?* Available at https://www.ei-ie.org/en/item/22632:whats-wrong-with-the-world-banks-human-capital-index-by-david-edwards

Galor, O., & Weil, D. N. (2000). Population, technology, and growth: From malthusian stagnation to the demographic transition and beyond. *American Economic Review, 90*(4), 806–828.

Goldin, C. (2016). Human capital. In C. Diebolt & M. Haupert (Eds.), *Handbook of cliometrics*. Springer Verlag.

Goldin, C., & Katz, L. F. (2024). The incubator of human capital: The NBER and the rise of the human capital paradigm, NBER chapters in *The economic history of American inequality: New evidence and perspectives.* National Bureau of Economic Research.

Goode, R. B. (1959). Adding to the stock of human and physical capital. *The American Economic Review, 49*(2), 147–155.

Hatch, N. W., & Dyer, J. H. (2004). Human capital and learning as a source of sustainable competitive advantage. *Strategic Management Journal, 25,* 1155–1178.

Hitt, M. A., Bierman, L., Shimizu, K., & Kochhar, R. (2001). Direct and moderating effects of human capital on strategy and performance in professional service firms: A resource-based perspective. *Academy of Management Journal, 44*(1), 13–28.

Huselid, M. A., Jackson, S. E., & Schuler, R. S. (1997). Technical and strategic human resource management effectiveness as determinants of firm performance. *Academy of Management Journal, 40*(1), 171–188.

Hyslop-Margison, E., & Sears, A. (2006). *Neoliberalism, globalization and human capital learning: Reclaiming education for democratic citizenship.* Springer.

Jiang, K., Lepak, D. P., Hu, J., & Baer, J. C. (2012). How does human resource management influence organizational outcomes? A meta-analytic investigation of mediating mechanisms. *Academy of Management Journal, 55*(6), 1264–1294.

Kiker, B. (1966). The historical roots of the concept of human capital. *The Journal of Political Economy, 74,* 481–499.

Mankiw, G., Romer, D., & Weil, D. N. (1992). A contribution to the empirics of economic growth. *Quarterly Journal of Economics, 107,* 407–437.

Mincer, J. (1958). Investment in human capital and the personal income distribution. *Journal of Political Economy, 66,* 281–302.

Mincer, J. (1974). *Schooling, experience, and earnings.* National Bureau of Economic Research.

Nussbaum, M. C. C. (2010). *Not for profits: Why democracy needs the humanities.* Princeton University Press.

OECD. (2001). *The well-being of nations: The role of human and social capital.* OECD.

Ployhart, R. E., & Moliterno, T. P. (2011). Emergence of the human capital resource: A multi-level model. *Academy of Management Review, 36*(1), 127–150.

Ployhart, R. E., Nyberg, A. J., Reilly, G., & Maltarich, M. A. (2014). Human capital is dead; Long live human capital resources! *Journal of Management, 40*(2), 371–398.

Ray, C., Essman, S., Nyberg, A. J., Ployhart, R. E., & Hale, D. (2023). Human capital resources: Reviewing the first decade and establishing a foundation for future research. *Journal of Management, 49*(1), 280–324. https://doi.org/10.1177/01492063221085912

Ridhwan, M. M., Nijkamp, P., Ismail, A., & Luthfi, M. I. (2022). The effect of health on economic growth: A meta-regression analysis. *Empirical Economics, 63,* 3211–3251. https://doi.org/10.1007/s00181-022-02226-4

Rosen, S. (1989). Human capital. In J. Eatwell, M. Milgate, & P. Newman (Eds.), *Social economics.*

The New Palgrave. Palgrave Macmillan. https://doi.org/10.1007/978-1-349-19806-1_19

Rumelt, R. P., Schendel, D. E., & Teece, D. J. (1994). Fundamental issues in strategy: A research agenda. In R. P. Rumelt, D. E. Schendel, & D. J. Teece (Eds.), *Fundamental issues in strategy: A research agenda for the 1990s.* Harvard Business School Press.

Schultz, T. W. (1961). Investment in human capital. *The American Economic Review, 51*(1), 1–17.

Schultz, W. (1972). Human capital: Policy issues and research opportunities. In T. W. Schultz (Ed.), *Economic research: Retrospect and prospect, vol 6: Human resources.* UMI.

Schultz, T. W. (1993). The economic importance of human capital in modernization. *Education Economics, 1*(1), 13–19.

Sevcenko, V., & Ethiraj, S. (2018). How do firms appropriate value from employees with transferable skills? A study of the appropriation puzzle in actively managed mutual funds. *Organization Science, 29*(5), 775–795. https://doi.org/10.1287/orsc.2017.1197

Sharpe, A. (2001). *The development indicators for human capital sustainability.* Centre for the Study of Living Standards, Canada. Available at http://www.csls.ca/events/cea01/sharpe.pdf

Skaggs, B. C., & Youndt, M. (2004). Strategic positioning, human capital, and performance in service organizations: A customer interaction approach. *Strategic Management Journal, 25,* 85–99.

Smith, A. ([1776] 2009). *An inquiry into the nature and the causes of the wealth of nations.* In C. Muir, & D. Widger (Eds.). Project Gutenberg.

Solow, R. M. (1956). A contribution to the theory of economic growth. *The Quarterly Journal of Economics, 70*(1), 65–94.

Solow, R. (1957). Technical change and the aggregate production function. *Review of Economics and Statistics, 39,* 312–320.

Somaya, D., Williamson, I. O., & Lorinkova, N. (2008). Gone but not lost: The different performance impacts of employee mobility between cooperators versus competitors. *Academy of Management Journal, 51*(5), 936–953.

Sweetland, S. R. (1996). Human capital theory: Foundations of a field of enquiry. *Review of Educational Research, 66*(3), 341–359. https://doi.org/10.3102/00346543066003341

Tan, E. (2014). Human capital theory: A holistic criticism. *Review of Educational Research, 84*(3), 411–445. https://doi.org/10.3102/0034654314532696

Teixeira, P. N. (2014). Gary Becker's early work on human capital – Collaborations and distinctiveness. *IZA Journal of Labor Economics, 3,* 12. https://doi.org/10.1186/s40172-014-0012-2

Unger, J. M., Rauch, A., Frese, M., & Rosenbusch, N. (2011). Human capital and entrepreneurial success: A meta-analytical review. *Journal of Business Venturing, 26,* 341–358.

Walsh, J. R. (1935). Capital concept applied to man. *The Quarterly Journal of Economics, 49*(2), 255–285.

WEF. (2015). *The human capital report: Employment, skills and human capital; Global challenge insight report.* A Publication of World Economic Forum (WEF).

White, L. H. (2017). *Human capital and its critics: Gary Becker, institutionalism, and anti-neoliberalism (January 25, 2017).* GMU working paper in economics no. 17-02, available at SSRN: https://ssrn.com/abstract=2905931 or https://doi.org/10.2139/ssrn.2905931

Winterton, J., & Cafferkey, K. (2019). Revisiting human capital theory: Progress and prospects. In K. Townsend, K. Cafferkey, A. M. McDermott, & T. Dundon (Eds.), *Theories of human resources and employment relations* (pp. 218–234). Edgar Publishing. https://doi.org/10.4337/9781786439017

World Bank. (2019). *The human capital project: Frequently asked questions.* Available at https://www.worldbank.org/en/publication/human-capital/brief/the-human-capital-project-frequently-asked-questions#:~:text=Human%20capital%20consists%20of%20the,as%20productive%20members%20of%20society

Wright, P. M., & McMahan, G. C. (2011). Exploring human capital: Putting human back into strategic human resource management. *Human Resource Management Journal, 21*(2), 93–104.

Wright, P. M., Coff, R., & Moliterno, T. P. (2014). Strategic human capital: Crossing the great divide. *Strategic Management Journal, 40*(2), 353–370.

Youndt, M. A., & Snell, S. A. (2004). Human resource configurations, intellectual capital, and organizational performance. *Journal of Managerial Studies, 16*(3), 337–360.

The Nature of Human Capital

3

Abstract

This chapter takes a look at the nature of human capital by examining its major components, sources, levels, and types of human capital using the knowledge-based conceptual framework. The framework relies principally on resource-based theory (RBT) and uses knowledge as a strategic resource. There are many other useful theoretical approaches and perspectives to understanding human capital. However, this book utilizes the lens of RBT to understand how countries, specifically African countries, can create value from human capital. The various specificity forms of human capital are better understood through the lens of the different dimensions of knowledge. This chapter helps to lay a better understanding of the nature of human capital after having defined the concept and clarified its meaning in the previous chapter.

Keywords

Human capital · KSAOs · Knowledge dimensions · Emergence · RBT

Introduction

One key to thoroughly understanding human capital is through examining its nature. Human capital is any attribute of individuals that can facilitate the creation of economic value one way or the other. Therefore, even though human capital cannot be seen, once an individual's attribute can facilitate production, it can be regarded as human capital. However, since the attributes of individuals that can be regarded as human capital depend on time, context, and space, human capital is dynamic. This is because individuals' attributes capable of creating economic value are dynamic. They are not static. They change with time and vary across contexts and space. Technology, for example, has rendered obsolete some skills that were once so valuable and highly sought after. There was a time the skills of being able to write in the lettering of "shorthand" and using the typewriter were so valuable that they were basic requirements for anyone to work as a secretary in an office or even to do some basic administrative work in the office. Currently, shorthand and typewriting skills are no longer as valuable in offices as they were once were. In fact, the typewriter machine has become obsolete and has been replaced by the keyboard of laptops and desktop computers.

Some skills and expertise are also more valuable in some firms and geographical contexts than in others. The dynamism of human capital also ensures that it can be acquired, developed, and enhanced with use. Unfortunately, it can also be depreciated with lack of use as the interplay of time, context, and space continually combines to affect its use or disuse. Human capital is clearly a multidimensional and an intangible construct. These two factors coupled together also make it a bit complex. Therefore, in order to understand its nature, its different dimensions and components need to be examined. A look at the sources, levels, and types of human capital enables an examination of its different dimensions and components. It also yields fruitful insights into understanding the value and potentials of human capital and how it can be a resource to individuals and groups.

The Main Components of Human Capital

From the definitions of human capital, its major components can be identified as knowledge, skills, and abilities (KSA). Knowledge, in particular, is a foundational component of human capital (Nyberg et al., 2014). Skills, abilities, and other human attributes capable of creating economic value are actually aspects of knowledge because knowledge can either be explicit or coded knowledge and tacit knowledge. Both explicit knowledge, knowledge that can be expressed and written which is why it is sometimes referred to as coded knowledge, and tacit knowledge, knowledge that comes from doing and applying explicit knowledge and which is sometimes referred to and translated to skills, are the foundational components of human capital.

In a knowledge economy, knowledge becomes even more important as the major element of human capital. Other components of human capital include diverse abilities and other characteristics of individuals (AOs) such as attitudes, talents, competences, ideas, information, health, and any ability that enables production and creates eco-

nomic value.[1] The inclusion of health as part of KSAOs deserves some special mention and clarification. On a cursory look, health may not easily come across as a component of human capital. However, at the very least, the implicit assumption is that for other KSAOs of individuals to have the capacity of producing economic value, individuals have to be healthy. Becker (2002) and Schultz (1993) expressly acknowledge the importance of health as a component of human capital by including it in their definitions. Individuals need to be healthy to be able to produce economic value. In other words, without good health, individuals will not be able to put their other KSAOs to productive use. The importance and economic value of health is confirmed by research which have long established that the differences in health of a population account for substantial differences in economic growth of countries (Lorentzen et al., 2008; Weil, 2007).

Still, at a deeper analysis, it is also conceivable and possible for some health conditions that ordinarily can be regarded as negative and consequently of no economic value to be a source of value. For instance, there are occasions where poor health conditions may give individuals the capacity to produce economic value. One may, for example, be able to give motivational talks and therefore be creating some economic value from such motivational talks because of personalized experiences from being in poor health. Such tacit knowledge gained becomes useful in creating some form of economic value. Obviously, these are extreme outlier examples of value from ill health. The more general case of health as human capital is good health.

Levels and Sources of Human Capital

At the very basic level, human capital can be at the intraindividual level and refers to specific components and attributes of individuals that

---

[1]Please see Table 2.1 in Chap. 2 to see various attributes of individuals included in its various definitions.

have economic value (Esho & Verhoef, 2020; Ployhart et al., 2014). At this level, reference is to specific knowledge, skills, abilities, and other attributes of an individual (KSAOs). Human capital can also be at the individual level such that each individual has a combination of attributes of KSAOs that have economic value. Lastly, human capital can also be at the group or collective level. At collective levels, human capital can be at the unit within an organization, firm or organizational level, or more collective levels relating to a society or other geographical spaces. Consequently, at the collective level, human capital is the sum that results from the accumulation or aggregation of the KSAOs of individuals. The summation of individuals' human capital can also result in collective capabilities. Therefore, wherever individuals are as a collective, there can be collective human capital and collective capabilities that become an attribute of the group. It is therefore not unusual to refer to human capital of firms, countries or nations, and even of units within firms. Each of these contexts is an organization and organization of individuals.

Recent research on strategic human capital by strategy scholars distinguish the concept of human capital resource (HCR) from human capital. This is partly an attempt to clarify the aggregate-level human capital that is available to firms as a resource, from individual levels of human capital that may not be available to firms (Moliterno & Nyberg, 2019).[2] The basic form of human capital belongs to the individual as all human capital is fundamentally embodied in individuals, and freewill ensures that it is individuals that decide the human capital they make available for use in their firms, countries, or any organizational context (Esho & Verhoef, 2020; Wright et al., 2001). A key fact is that human capital is part of the individual person and cannot truly be separated from the person.

However, human capital at the collective level is not simply a summation by addition of the human capital of individuals in that space. No.

For human capital to be a resource at the collective level, the collective human capital has to be the result of a systematic process. Ployhart and Moliterno (2011) refer to this process, in the context of a firm, as an emergence process. Group or collective human capital emerges from the interaction of the KSAOs of individuals in a space interacting with the various contextual dimensions in that space. Collective human capital also depends on the various quantity and quality of KSAOs that comprise the collective unit or space. Different forms of unit-level human capital may also emerge from the "emergence enabling process" (Ployhart & Moliterno, 2011: p. 128). In the same vein, Ployhart et al. (2014) differentiate between human capital and human capital resource by arguing that human capital resources, at individual and macro levels such as units and firms, are created from human capital combinations through the processes of emergence and complementarity. Therefore, there are different forms of human capital resources that emerge through the process of emergence and complementarity when human capital at lower levels is combined. For clarity, an individual with various KSAOs has different forms of human capital resources that can emerge when the individual attributes with productive value are combined. There is also, simultaneously, a form of complementarity that emerges with each combination such that each combination is capable of producing an economic value higher than what would have been possible without the combination. Consequently, also at the collective level, for human capital to result in a collective capability and become human capital resources, there has to be an effective emergence process that results in complementarities. Collective human capital resource requires an effective emergence process to culminate into a holistic whole and have complementarity effects. This process, if well managed, ensues that the collective human capital resource is greater than the sum of its constituent parts. In other words, the collective human capital at a group level, if well managed, becomes collectively greater than the combinations of the human capital of the individuals in the group. Summarily, at different levels of human capital,

---

[2] Human capital and human capital resources are regarded as one and the same in this chapter and throughout the book for consistency.

individual, unit, firm, or higher organizational levels such as societies and countries, an effective emergence process with complementarity effects increases productivity levels.

Like all other forms of capital, the acquisition and accumulation of human capital requires investment. There are diverse means through human capital can be acquired, developed, and accumulated, and these can be regarded as the sources of human capital. Indeed, the sources of human capital are almost unlimited. Investments into acquiring and developing any individual attribute that has productive value result in increasing the human capital of the individual. This can subsequently translate into the human capital of the collective in that organizational space. However, the nature of the resulting collective human capital depends on many factors and circumstances within that particular organizational context of the group.

Sources of Human Capital at the Individual Level As the foundational component of human capital, knowledge can be derived from education and from work experience (Schultz, 1993; Nyberg et al., 2014). Although when it comes to human capital, the immediate focus is usually knowledge that comes from formal education, knowledge can be derived from any form of education and training, formal or informal. An often-forgotten source of education and knowledge, and by extension human capital, is informal training that is derived from the home and family. Other than the formal school system that usually consists of primary, secondary, and tertiary levels of education, formal apprenticeship programs that equip individuals with specific skills that come majorly by doing and hands-on approach are also great sources of human capital. Personal development training such as formal workshops and seminars can also be sources of individual human capital. While investment in formal education is one main source of human capital, another major source of human capital is on-the-job training and workplace experience. It is important to understand that any system or scheme that enables skills acquisition and imparts abilities is a source of human capital.

Consequently, human capital is not limited to "western styles or forms of education." Unbeknownst to many people, investments in health constitute an important, yet often unrecognized, source of human capital which should never be taken for granted.

Sources of Collective Human Capital The major source of human capital for organizations at the firm or larger collective levels is the individual. For a firm or unit in an organization, hiring individuals from other firms or from the labor market is one major source of human capital. However, firms may also access and develop their human capital through mergers, acquisitions, and strategic alliances (Esho & Verhoef, 2020; Ferrary, 2015; Ranft & Lord, 2000). Through these processes, firms can access the knowledge base and other forms of human capital from individuals and also develop these individuals through on- and off-the-job training to simultaneously increase the individual's human capital and their own collective human capital. Investments in health programs and insurance for employees invariably constitute investments in human capital although these are not as direct and as explicit as other investments and sources. The productive capacity of firms can be seriously hampered by poor employees' health. At a more aggregate level such as countries, immigration can be a source of human capital. In fact, a realization that immigration can be a source of human capital caused Schultz (1961) to include migration as one of the five main categories of human capital in his early categorization of the concept; the other four categories are health, on-the-job training, schooling, and adult education. Coordinated international immigration programs for countries such as those of Canada, Australia, the United Kingdom (UK), and the United States (USA) are sometimes targeted to source for human capital—specific forms of human capital in certain subject or knowledge areas. Consequently, organized immigration programs are beyond individuals changing geographical location. When assessing a country's stock of human capital, it needs to be recognized that immigration and emigration are, respectively, additions and losses to a country's stock of human capital (Abraham & Mallatt, 2022). Issues of immigration can be issues of human capital just as lateral mobility from one firm to the other are transfers and movements of human capital (Esho & Verhoef, 2020).

Types of Human Capital

Human capital may be classified into a number of categories using different factors as a means of categorization. It can, for example, be categorized according to their sources stated in the previous section. Consequently, it is possible to classify human capital into health human capital, education or knowledge human capital, skills human capital, and so forth. However, the classification used here will be that derived from analyzing the foundational component of human capital—knowledge. Using the many sources of human capital to categorize human capital runs the risk of miniaturing its types into diverse classes. It also gives the impression that each type can stand and create economic value alone. In contrast, human capital though multidimensional is a complex whole that creates and produces economic value in unison with other components from different sources. This fact is not always obvious because one main component from a unique source may be the primary producer of value in a given economic activity at one point in time. For example, the acquisition of data analytic skills from formal education may be used primarily in a job role. However, health, other general knowledge, and other skills derived from other sources are simultaneously at work, while data analytic skills are being primarily utilized in the job role.

Human capital can thus be classified using its different dimensions into various types. To classify human capital, a conceptual framework developed from the different dimensions of knowledge is adopted because it aligns and fits with Becker's (1962, 1964) categorization that differentiates general human capital from firm-specific human capital. It also fits with more recent nuances that categorize human capital into

industry- (e.g., Kim et al., 2014; Neal, 1995; Parent, 2000), occupation- (Groen, 2006; Kambourov & Manovskii, 2009; Zangelidis, 2008), country- (Kim & Park, 2013; Priyo, 2012), and task-specific human capital (e.g., Gathmann & Schonberg, 2010; Yamaguchi, 2012). The knowledge-based conceptual framework below also accommodates these recent nuanced categorizations.

A Knowledge-Based Conceptual Framework of Human Capital

The framework draws from resource-based theory (RBT) and knowledge-based view of the firm (KBV). The major proposition of RBT is that strategic firm resources are valuable, rare, difficult to imitate, and non-substitutable (VRIN) (Barney, 1991), and competitive advantage and superior performance comes from acquisition and deployment of strategic resources. Clearly, RBT is a firm-level theory and applies largely to firms. So, through the RBT lens, human capital can be a strategic resource if each of the KSAOs of the individual or KSAOs in combination has VRIN qualities. However, what about the case of countries? Does and can the logic of RBT apply to countries? The third part of this book uses the logic of RBT to present ideas on how African countries can make more strategic investments in and create value from human capital.

In KBV, knowledge resides principally in individuals, and the role of the firm rests in the integration of this knowledge for productive activities (Grant, 1996a, b). In addition to utilizing knowledge as residing primarily in individuals, knowledge is also recognized as a strategic resource of the firm capable of aiding the firm to attain competitive advantage and superior performance. Although the primary locus of knowledge was hitherto the subject of much debate in the knowledge literatures (Felin & Hesterly, 2007), the literature seems, at least, to be in agreement either by explicit declarations or implication that knowledge resides in individuals as well as firms (Kogut & Zander, 1992; Nonaka, 1994; Spender, 1996). This knowledge that resides in individuals forms the foundation of human capital (Nyberg et al., 2014), and individuals are ultimately the owners of their human capital (Esho & Verhoef, 2020). Consequently, examining the different dimensions of knowledge helps to shed light on the different types of human capital and in understanding the nature of human capital generally, apart from its characteristics as a form of capital.

Dimensions of Knowledge

Knowledge may be classified along three different dimensions: its degree of transferability across individuals, its degree of transferability across organizational contexts, and its level of generality and specialty across different knowledge domains. Using these three main dimensions, different types of human capital emerge.

Dimension 1: Degree of Transferability Across Individuals Under this dimension, knowledge can be classified as either tacit or explicit. The tacit–explicit dimension is in relation to the degree to which knowledge can be transferred or communicated between persons (Grant, 1996a; Nonaka, 1994). Explicit knowledge is also referred to as codified knowledge, and it is that aspect of knowledge that can be transmitted or communicated in formal systematic language (Nonaka, 1994). Tacit knowledge, on the other hand, is that aspect of knowledge that is difficult to communicate even though the individual may be able to apply that knowledge in performing a task. It is that form of knowledge aptly expressed in the now famous words of Michael Polanyi "we can know more than we can tell." Tacit knowledge is difficult to express even though it can be practicalized. Explicit knowledge is associated with knowing "what," while tacit knowledge basically refers to knowing "how." This, therefore, implies that it is possible to know "what" without knowing "how" and to know "how" without exactly knowing how to express the "how" in definite formal language. Consequently, the possibility of also knowing "how" without a

solid foundation of the "what" behind the "how" also exists and vice versa.

Tacit knowledge may be further divided into its cognitive and technical elements (Nonaka, 1994). However, the distinction between explicit and tacit knowledge serves the purpose of helping to understand levels of human capital. From this distinction, human capital may be skills-based or non–skills-based. This raises the question of the meaning of skills itself. There are various definitions of skills, but the one that best aligns with our purpose here is that which defines skills as the "expertise that has been developed through training and experience" (Zhang, 2019: p. 1). Vocational education and training, apprenticeships, internships, and training programs that educate through learning-by-doing equip individuals with skills-based human capital. Non-skills human capital, on the other hand, are forms of human capital that are developed through transmission and receipt of express or codified knowledge. Consequently, skills-based human capital can be seen as more directly practical than non–skills-based human capital. However, the dichotomy between these two types of human capital is usually more nuanced in practice. Indeed, in many instances, these two types of human capital can hardly be strictly differentiated, although in some instances, they can. The backbone of skills-based human capital is usually a set of codified or express knowledge that is communicated before experiential or tacit knowledge begins.

Dimension 2: Degree of Transferability Across Organizational/Collective Contexts Knowledge can also be general or specific to certain organizational contexts depending on the extent to which persons can transfer it across these contexts. Using this second dimension, organizational context could range from firms or formal organizations to industries, occupations, societies, and countries. When knowledge is transferable across organizational contexts, the implication is that it can be applied across contexts. Knowledge that cannot be transferred across contexts implies that it cannot be applied across contexts. Becker's (1962, 1964) categorization relied on the degree to which human capital can be applied within a firm. Consequently, his distinction between firm-specific and general human capital stemmed from this recognition that there are types of human capital that can only be applied in one firm and others that are more general and therefore can be applied across firms. However, many thanks to recent economics research, human capital may also be occupation-specific (Groen, 2006; Kambourov & Manovskii, 2009; Robinson, 2018; Zangelidis, 2008), industry-specific (Kim, 1992; Kim et al., 2014; Lagoa & Suleman, 2016; Neal, 1995; Parent, 2000), country-specific (Adda et al., 2022; Kim & Park, 2013; Priyo, 2012), and task-specific (Esho & Adegbesan, 2024; Esho & Verhoef, 2020; Gibbons & Waldman, 2004; Gathmann & Schonberg, 2010; Yamaguchi, 2012). Implied in these nuanced forms of human capital is the degree of its transferability and application of different forms of knowledge to different contexts. Clearly, this dimension has been the mostly recognized in extant literature. Consequently, several studies have been conducted using these various forms to investigate wage costs and differentials, compensation, workers' mobility, performance, human resource configuration, organizational learning modes, and other issues.

Each of these two categorizations speaks to different types knowledge and consequently what can be regarded as different types of human capital. Extant literature in management and economics has utilized these dimensions of knowledge to build theories and frameworks with great utility. Knowledge management stream, beginning from Nonaka (1994), for example, utilized the explicit–tacit dimension of knowledge to theorize the four modes of knowledge creation: socialization, combination, externalization, and internalization. Classifying knowledge along the dimension of specificity and transferability to an organizational context, Becker (1962, 1964) distinguished firm-specific human capital from general human capital. Although he did not expressly acknowledge this dimension of knowledge in his classification, it was implicit in his classification and analysis.

While Becker's classification was more directly related to human capital than Nonaka's (1994) usage, their usage of these dimensions has helped to reveal tremendous insights and is a pointer to the insights that could be gleaned when human capital is viewed from the dimensions of its major component—knowledge.

Dimension 3: Specialized-General Knowledge - Specialist and Generalist Human Capital Using this dimension, human capital may be specialist or generalist. Specialists have knowledge in a particular knowledge domain, while generalists have knowledge that may cut across different knowledge domains or disciplines. Specialists are usually experts and have deeper but narrower scope of knowledge in a particular domain. The time required to become a specialist is usually longer than that required to become a generalist. Generalists, on the other hand, have a broader scope of knowledge and possess a macro-orientation unlike specialists who are more interested in deepening their knowledge and often narrowing their knowledge in a single discipline or subject area and perfecting their knowledge in that area in the process (Cesare & Thornton, 1993). Professionals who, most times, belong to professional groups attained through certification are good examples of specialists. However, generalist or specialist knowledge is not only attained through formal education. Either one of them can also be derived from work experience. Generalist human capital can result from the accumulation of broad career experience across industries, while specialist human capital can be that acquired from a single firm or industry (Agnihotri & Bhattacharya, 2021; Li & Patel, 2019; Mueller et al., 2021). Between the two extremes of generalists and specialists are diverse combinations involving different levels of specialty. This dimension has found more usage and acceptance in practice than the previous two dimensions for obvious reasons. Practitioners can more directly relate to this form of specificity than other forms of specificity. Some degrees of usage can be also found in many professions such as law, accounting, finance, and the medical and health professions. Consequently, this generalist–specialist specificity is more readily acceptable and in use among practitioners.

The different types of human capital derived from the knowledge-based framework are not mutually exclusive, and indeed, the categorizations are not intended to be. Therefore, one form of human capital may fit in neatly into more than one category. Moreover, in reality, individuals do not consciously split their human capital into different types before deployment in productive activities. Human capital is cumulatively accumulated over time with each new addition integrating holistically into an integrated complex bundle that cannot be really totally unbundled into its different constituent parts.

Studies in human resource management, strategic management, and management field in general have tended to approach all human capital specificity as one and the same. The specialist versus generalists dimension can hardly be differentiated from the firm specificity–general human capital dimension. Although studies in these two dimensions have tended to be different, they have, however, not been any explicit demarcation between the two types of specificity belonging to different, though somewhat related, dimensions. The framework in Fig. 3.1 separates these two dimensions while simultaneously acknowledging the relationships between the two stemming from the different knowledge dimensions. This knowledge-based conceptual framework aids in revealing the variations in human capital and promises some reward for future research.

The knowledge categorizations utilized in the framework are not the only dimensions of knowledge. These three main dimensions serve the purpose of helping to understand the nuances and intricacies in the nature of human capital, particularly in the types of human capital. Some research use a different approach and categorize human capital according to the function it performs such as founders' human capital (e.g., Criaco et al., 2014; Colombo & Grilli, 2005; Ganotakis, 2012), managerial human capital (e.g., Castanias & Helfat, 2001; Crocker & Eckardt, 2014; Zhao & Thompson, 2019), CEOs' human capital (e.g., Agnihotri & Bhattacharya,

2021; Li & Patel, 2019), entrepreneurial human capital (Krieger et al., 2022; Queiro, 2022), and so forth. These categorizations also have good utility as they enable one to immediately identify the major work activity to which human capital is being applied or deployed. However, the knowledge-based conceptual framework presented in this chapter is able to accommodate these and other modes of classifications as it can basically align with the degree to which human capital is applicable to a context, the main task to which human capital is being applied, or the specificity to an identified task (see Fig. 3.1). Unveiling the various types of human capital furthers the understanding of the nature of human capital at different levels and aids a better understanding of the outcomes of human capital, also at different levels.

The Emergence of Collective Human Capital: The Case for National Human Capital

The knowledge-based framework has the additional utility of extending beyond the firm context to any broader organizational context such as a community, society, or country. Just as firms can attain competitive advantage, countries can also achieve the same (Porter, 1990). A country may also be looked at through the lens of a broad organizational context based on geography and one that that helps individuals within their geography to utilize their human capital to productive use. Such human capital could also be steered toward helping the country gain a competitive advantage among nations. From this perspective, a country or any other geographical context, for that matter, is really not much different from a firm. Well, apart from some differences such as the firm being in business to make profits or achieve some other social objectives, it is not very different from a country. Individuals, like workers within a firm, aim to achieve good success and performance as the firms they work for also aim for competitive advantage and performance. Countries may have other loftier reasons for being a country, but they also aim to be productive just like firms.

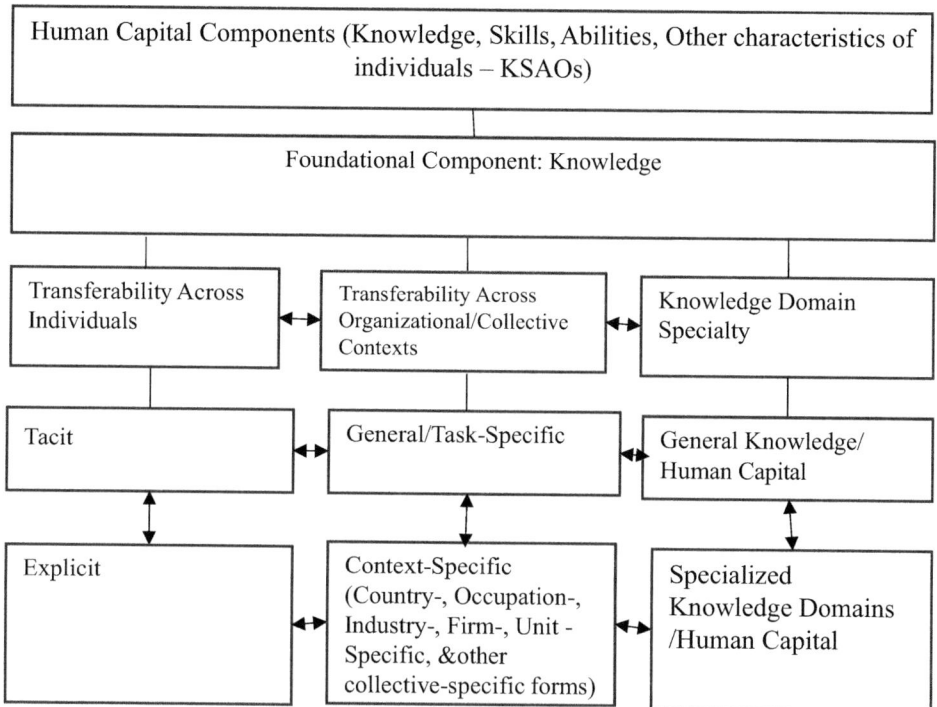

Fig. 3.1 The knowledge-based conceptual framework of human capital

National human capital can be said to be that human capital for which a country is best known for and which has the ability to help the country to attain a competitive advantage among countries. In this sense, national human capital is not merely equal to the notion of country-specific human capital, that is, human capital that is more useful in country than in others. National human capital is the output of an emergence process that can be similar to that described in Ployhart and Moliterno (2011). It can also be the output or result of a deliberate emergence process that ensures that national human capital is formed through a deliberate formation process rather than a mere accidental emergence process. Although, the main objective of the process may not be to end up with national human capital, one of the resultant output ends up being one nevertheless—a form of human capital that can be attributed to a particular country. In this regard, fortunately, African countries have a rich heritage of indigenous knowledge that is embedded in the diverse social and cultural heritage of African societies. These can be exploited toward the emergence of unique forms of national human capital.

The idea that a country can have a unique form of human capital may appear far-fetched. However, in reality, countries have been known for particular skills or capabilities. Italy has been known to be a producer of quality leather products, while France has been known for high fashion and for its fine cuisine. Japan emerged sometime in the 1970s and 1980s to become known as the producer of energy or fuel-efficient cars. Switzerland is globally renowned for its pharmaceuticals and choice chocolates and Germany for high-performing cars. The list could go on and on. With the exemption of natural resources or primary products, the emergence of a particular product from a country requires the utility of specific types of human capital. For one reason or the other, a country becomes well vest in producing a particular good or service. This, in most cases, is not accidental. Although the countries may not have intentionally set out to become known in a specific area, the resulting effect of becoming known did not occur in a vacuum.

Consequently, the notion of national human capital should not be as odd as it may have sounded at first read.

Conclusion

As a complex, multidimensional construct, human capital may be better understood through knowledge as its foundational component. Just as individuals can acquire and develop their human capital, organizational groups such as countries can also develop and accumulate human capital. Countries can become deliberate about enabling the development of specific capabilities based on their stock of human capital.

References

Abraham, & Mallatt. (2022). Measuring human capital. *Journal of Economic Perspectives, 36*(3), 103–130.

Adda, J., Dustmann, C., & Gorlach, J. (2022). The dynamics of return migration, human capital accumulation, and wage assimilation. *The Review of Economic Studies, 89*(6), 2841–2871.

Agnihotri, A., & Bhattacharya, S. (2021). Generalist versus specialist CEO and R&D commitment: Evidence from an emerging market. *Journal of Management & Organization*, 1–17. https://doi.org/10.1017/jmo.2021.7

Barney, J. (1991). Firm resources and sustained competitive advantage. *Journal of Management, 17*(1), 99–120.

Becker, G. S. (1962). Investment in human capital: A theoretical analysis. *The Journal of Political Economy, 70*(5), 9–49.

Becker, G. S. (1964). *Human Capital: A theoretical and empirical analysis, with special reference to education.* University of Chicago Press.

Becker, G. S. (2002). The age of human capital. In E. P. Lazear (Ed.), *Education in the twenty-first century* (pp. 3–8). Hoover Institution Press.

Castanias, R. P., & Helfat, C. E. (2001). The managerial rents model: Theory and empirical analysis. *Journal of Management, 27*, 661–678.

Cesare, S. J., & Thornton, C. (1993). Human resource management and the specialist/generalist issue. *Journal of Managerial Psychology, 8*(3), 31–40. https://doi.org/10.1108/02683949310027763

Colombo, M. G., & Grilli, L. (2005). Founders' human capital and the growth of new technology-based firms: A competence-based view. *Research Policy, 34*, 795–816.

Criaco, G., Minola, T., Migliorini, P., & Serarols-Tarrés, C. (2014). To have and have not': Founders' human capital and university start-up survival. *Journal of Technology Transfer, 39*, 567–593.

Crocker, A., & Eckardt, R. (2014). A multi-level investigation of individual- and unit-level human capital complementarities. *Journal of Management, 40*(2), 509–530.

Esho, E., & Adegbesan, T. (2024). *When does the specificity of human capital lead to superior performance?* Proceedings of the Academy of Management Conference. https://doi.org/10.5465/AMPROC.2024.19596abstract

Esho, E., & Verhoef, G. (2020). A holistic model of human capital for value creation and superior firm performance: The strategic factor market model. *Cogent Business & Management, 7*(1), 1728998. https://doi.org/10.1080/23311975.2020.1728998

Felin, T., & Hesterly, W. S. (2007). The knowledge-based view, nested heterogeneity, and new value creation: Philosophical considerations on the locus of knowledge. *Academy of Management Review, 32*(1), 195–218.

Ferrary, M. (2015). Investing in transferable strategic human capital through alliances in luxury hotel industry. *Journal of Knowledge Management, 19*(5), 1007–1028. https://doi.org/10.1108/JKM-01-2015-0045

Ganotakis, P. (2012). Founders' human and the performance of UK new technology-based firms. *Small Business Economics, 39*, 495–515. https://doi.org/10.1007/s11187-010-9309-0

Gathmann, C., & Schonberg, U. (2010). How general is human capital? A tasks-based approach. *Journal of Labour Economics, 28*(1), 1–49.

Gibbons, R., & Waldman, M. (2004). Task-specific human capital. *American Economic Review, 94*(2), 202–207.

Grant, M. G. (1996a). Toward a knowledge-based theory of the firm. *Strategic Management Journal, 17*, 109–122.

Grant, M. G. (1996b). Prospering in dynamically-competitive environments: Organizational capability as knowledge integration. *Organization Science, 7*(4), 375–387.

Groen, J. A. (2006). Occupation-specific human capital and labour markets. *Oxford Economic Papers, 58*, 722–741.

Kambourov, G., & Manovskii, I. (2009). Occupational specificity of human capital. *International Economic Review, 50*(1), 63–115.

Kim, D. (1992). *Industry wage differences: The unobservable human capital hypotheses.* University of Chicago.

Kim, J., & Park, J. (2013). Foreign direct investment and country-specific human capital. *Economic Inquiry, 51*(1), 198–210.

Kim, K., Mithas, S., Whitaker, J., & Roy, P. K. (2014). Industry-specific human capital and wages: Evidence from the business process outsourcing industry. *Information Systems Research, 25*(3), 618–638.

Kogut, B., & Zander, U. (1992). Knowledge of the firm, combinative capabilities, and the replication of technology. *Organization Science, 3*(3), 383–397.

Krieger, A., Stuetzer, M., Obschonka, M., & Salmela-Aro, K. (2022). The growth of entrepreneurial human capital: Origins and development of skill variety. *Small Business Economics, 59*, 645–664.

Lagoa, S., & Suleman, F. (2016). Industry- and occupation-specific human capital: Evidence from displace workers. *International Journal of Manpower, 37*(1), 44–68.

Li, M., & Patel, P. C. (2019). Jack of all, master of all? CEO generalist experience and firm performance. *The Leadership Quarterly, 30*(3), 320–334.

Lorentzen, P., McMillan, J., & Wacziarg, R. (2008). Death and development. *Journal of Economic Growth, 13*(2), 81–124.

Moliterno, T. P., & Nyberg, A. J. (2019). Strategic human capital resources: A brief history, construct definition, and introduction to the handbook of research on strategic human capital resources. In *Handbook of research on strategic human capital resources* (pp. 2–12). Edward Elgar Publishing.

Mueller, P. E. M., Georgakakis, D., Greve, P., Peck, S., & Ruigrok, W. (2021). The curse of extremes: Generalist career experience and CEO initial compensation. *Journal of Management, 47*(8), 1977–2007. https://doi.org/10.1177/0149206320922308

Neal, D. (1995). Industry-specific human capital: Evidence from displaced workers. *Journal of Political Economy, 13*(4), 653–677.

Nonaka, I. (1994). A dynamic theory of organizational knowledge creation. *Organization Science, 5*(1), 14–37.

Nyberg, A. J., Moliterno, T. P., Hale, D., Jr., & Lepak, D. P. (2014). Resource-based perspectives on unit-level human capital: A review and integration. *Journal of Management, 40*(1), 316–346.

Parent, D. (2000). Industry-specific capital & the wage profile: Evidence from the National Longitudinal Survey of Youth and the Panel Study of Income Dynamics. *Journal of Labour Economics, 18*(2), 306–323.

Ployhart, R. E., & Moliterno, T. P. (2011). Emergence of the human capital resource: A multi-level model. *Academy of Management Review, 36*(1), 127–150.

Ployhart, R. E., Nyberg, A. J., Reilly, G., & Maltarich, M. A. (2014). Human capital is dead; Long live human capital resources! *Journal of Management, 40*(2), 371–398.

Porter, M. (1990). The competitive advantage of nations. *Harvard Business Review*, March–April, 73–91. Available at https://economie.ens.psl.eu/IMG/pdf/porter_1990_-_the_competitive_advantage_of_nations.pdf

Priyo, A. K. K. (2012). Sector-specific capital, labor market distortions and cross-country income differences: A two-sector general equilibrium approach. *The BE Journal of Macroeconomics, 12*(1), Article 3.

Queiro, F. (2022). Entrepreneurial human capital and firm dynamics. *The Review of Economic Studies, 89*(4), 2061–2100.

Ranft, A. L., & Lord, M. D. (2000). Acquiring new knowledge: The role of retaining human capital in acquisitions of high-tech firms. *The Journal of High Technology Management Research, 11*(2), 295–319.

Robinson, C. (2018). Occupation mobility, occupation distance, and specific human capital. *Journal of Human Resources, 53*(2), 513–551.

Schultz, T. W. (1961). Investment in human capital. *The American Economic Review, 51*(1), 1–17.

Schultz, T. W. (1993). The economic importance of human capital in modernization. *Education Economics, 1*(1), 13–19.

Spender, J. C. (1996). Managing knowledge the basis of a dynamic theory of the firm. *Strategic Management Journal, 17*, 45–62.

Weil, D. (2007). Accounting for the effect of health on economic growth. *Quarterly Journal of Economics, 122*(3), 1265–1306.

Wright, P. M., Dunford, B. B., & Snell, S. A. (2001). Human resources and the resource based view of the firm. *Journal of Management, 27*, 701–721.

Yamaguchi, S. (2012). Tasks and heterogeneous human capital. *Journal of Labour Economics, 30*(1), 1–53.

Zangelidis, A. (2008). Occupational and industry specificity of human capital in the British labour market. *Scottish Journal of Political Economy, 55*(4), 420–443.

Zhang, C.-Q. (2019). Skill. In D. Hackfort, R. J. Schinke, & B. Strauss (Eds.), *Dictionary of sport psychology: Sport, exercise, and performing arts* (pp. 270–272). Academic Press.

Zhao, Y., & Thompson, P. (2019). Investments in managerial human capital: Explanations from prospect and regulatory focus theories. *International Small Business Journal, 37*(4), 365. https://doi.org/10.1177/0266242619828264

Outcomes and Benefits of Human Capital

4

Abstract

This chapter focused the discussion on the various benefits of human capital to individual persons and groups. The outcomes of human capital have been presented through the multidisciplinary lenses of human resource (HR) management, strategy and strategic management, and economics. Human capital increases wages and earning potentials of individuals and increases the performance of business firms and all forms of organizations. In formal organizations, HR practices constitute a form of institutions that guide the management of people with human capital to ensure that the knowledge, skills, abilities, and other characteristics they possess are utilized for the optimum productivity of their organizations. Societies also benefit from the accumulation of human capital by persons. Although human capital outcomes are sometimes direct, they are more complex in societies simply because societies are multifaceted. This chapter concludes with the "micro–macro paradox" on the returns to human capital and introduces the Human Capital Returns Matrix.

Keywords

HR practices · Micro–macro paradox · Individual outcomes · Collective outcomes · Human capital

Introduction

The outcomes of human capital refer to its effects and impacts on individuals, work, business and organizations, and societies, up to country levels. These human capital outcomes are many and varied. While some outcomes are direct, others are indirect. Indirect human capital outcomes may not be as obvious as the direct outcomes, but they result from the accumulation of human capital nonetheless. Human capital outcomes go beyond economic outcomes to cut across social and cultural spheres of a place, to the psychological and health outcomes of individuals and nations. Indeed, wherever an aggregation of individuals can be congregated semipermanently, the impact of the human capital in that space can usually be felt, especially if the available human capital is of satisfactory or right quality. However, it is impossible to fully mention and discuss all the possible human capital outcomes. Therefore, this chapter presents the major outcomes that have been proven beyond doubt in much of extant research in various academic fields. The empirical evidence of these outcomes is also beyond doubt, and they help to give an idea of the possibilities inherent in and from human capital as a resource to individuals and collective groups.

An alternative word that could be used to replace outcomes is benefits. However, using the word benefits necessitates the additional responsibility of adding an object—benefits to whom?

© The Author(s), under exclusive license to Springer Nature Switzerland AG 2024
E. Aziegbe-Esho, *On the Sustainable Development of African Countries*, Sustainable Development Goals Series, https://doi.org/10.1007/978-3-031-81124-1_4

It would become necessary to state the persons that benefit from human capital. Individuals with human capital and the collectives or groups to which they belong all benefit from investment and accumulation of human capital. In other words, it is not only the specific individual with human capital that benefits. The organization(s) and society at large to which the individual belongs derive enormous direct and indirect benefits. In the following sections, the outcomes of human capital at different levels are presented with the primary beneficiaries of the benefits stated in the subheadings.

Human Capital Outcomes at the Individual Level (Benefits to Individual Owners of Human Capital)

One of the earliest established outcomes of individuals' human capital is increased wages and earning potential (Becker, 1964). Returns to formal schooling and years of formal education constitute the earliest forms of research into human capital. In a sense, human capital is a concept in which the returns and outcomes were studied long before the formal crystallization, recognition, and acceptance of the concept itself. Indeed, attempts to establish that the returns to investments in formal education are greater than its costs constitute the very bedrock of the earliest studies in the concept. In addition to Berker's research works on the concept, Walsh's (1935), Schultz's (1961), and Mincer's (1974) interests in the returns to formal schooling and professional training formed part of early studies on human capital. The numerous studies on the returns to formal education that now abound have gone beyond trying to establish the concept of human capital to nuances on the nature of these earnings. Other aspects of early research on returns to schooling have spun into various strands of research that are not directly in tangent with mainstream human capital research. Research on returns to schooling still exists albeit with various other foci (Teixeira, 2014).

Returns to individuals on their human capital include increased wages, salaries, other monetary benefits, and other forms of increased earnings such as promotion to higher levels in the workplace. Different types of human capital will command different levels of wages and earnings. Generally, specialized and specific forms of human capital may lead to higher wages and earnings. Just as human capital varies between firms, occupations, industries, and other contexts, the wages accruing to individuals with human capital also vary according to job, geography, and other contexts. Consequently, there are industry-, occupation-, country-, and firm-specific wage differentials. Earning and wage outcomes to an individuals' human capital will vary across these different contexts.

In addition to increased earnings, individuals with human capital, all things being equal, perform better on their job. Increased job performance is a fundamental outcome of human capital. Theoretically, it is the implicitly expected increase in job performance that actually leads to the increase in wages and potential earnings. The positive returns to schooling and on-the-job or professional training, forms of general and specific human capital, respectively, result from expectations of higher job performance. Although actual performance may eventually differ from expectations, job performance and increased earnings are joint twin outcomes of human capital at the individual level. While there are situations where this may not necessarily be the case, these cases are exceptions, and other factors such as motivation and commitment may be the contributing factors of such situations. A worker that is employed or promoted because of social relationships with management without having the requisite skills to perform on the job, for example, may perform below expectation despite the increased earnings and wages accruing to the worker.

Research linking human capital to job performance and human capital to wages have been conducted in divergent fields. On one hand, human resource management (HRM) research have tended to concentrate more on the management of humans or people as resources by HRM departments or units within organizations. Therefore, the focus is usually on human resource practices, policies, and systems. On another hand, the related field of organization behavior (OB) has focused on other aspects of persons that work within organizations. These two fields, HRM and OB, while offering invaluable insights to the management of people and behavior within organizations have inadvertently paid little attention to the relationship between human capital itself and job performance. Although there are studies of the effects of human capital on job performance, they seem to be disjointed and do not form a cohesive body of research. Relatedly, studies linking human capital to wages are clearly denominated within economics. The resultant cumulative effect of these divergent academic disciplines and non-cohesive body of research is that studies relating human capital to job performance and to wages have rarely "spoken" to one another or been integrated. Consequently, the implicit link between job performance expectations and increased wages has not been made obvious, although the link appears intuitively obvious, and Becker's (1964) analysis had brought this implicit link to the fore albeit indirectly.[1]

Another important outcome of human capital for individuals is innovation and entrepreneurship ability. Human capital increases individuals' abilities to discover and exploit business opportunities. Indeed, empirical studies have found that generally, human capital is linked to success in entrepreneurship (Dimov, 2017; Martin et al., 2013; Unger et al., 2011). This positive link between human capital and entrepreneurship success is not limited to specific entrepreneurship education and training that is fast becoming common in universities; all forms of human capital, not just that acquired through formal education, increase the chances for succeeding as an entre-

preneur. This accounts for why venture capitalists typically use human capital attributes such as management skills and experience as their selection criteria for new investment ventures (Zacharakis & Meyer, 2000).

Human capital improves the quality of life, health, and general well-being (Becker, 2007; Bloom & Canning, 2003; Filmer et al., 2021). In fact, health itself is a form of human capital that is the underlying enabling factor of other forms of human capital. Indeed, health "increases the future productive power of individuals and the economy" (Bloom & Canning, 2003: p. 305). The psychological and social benefits of human capital to individuals are numerous and enormous although this is not always so obvious. People's employability usually increases with the acquisition of different types and higher levels of human capital. This is why, for the most part, college, university graduates, and certified professionals in some occupations have the potential to get higher-paying jobs than those without these qualifications. Essentially, modern societies have come to accept the acquisition of formal education and schooling as the normal requirement for increasing one's opportunities at becoming gainfully employed. This has become even more so in the knowledge economy. Given the historical trajectory of the knowledge economy and the rapid advent and advancement of new technologies, the knowledge economy has come to stay. The emerging Fourth Industrial Revolution (4IR)[2] is not expected to reverse the knowledge economy although there might be changes in the type of human capital required. Rather, the new emerging digitalized economy builds on the foundations of the knowledge economy.

Summarily, human capital equips individuals with capacity to innovate, increase their earnings through work especially knowledge work, perform better on their jobs in organizations and

---

[1]Boon et al.'s (2019) important call to integrate human capital research in SHRM and strategy arose out of the disjointed study of human capital. Multi- and interdisciplinary research are also needed to further integrate studies of human capital across other disciplines where it is possible to do so given that there can be divergent research interests in disciplines.

[2]4IR was coined by the founder of World Economic Forum (WEF), Klaus Schwab. Some do not agree the world is entering a Fourth Industrial Revolution (e.g., Rifkin, 2012, 2016; Moll, 2021). Nevertheless, artificial intelligence (AI), robotics, 3D and 4D printing, and similar technologies indicate that the technological advancements are entering a whole new level, 4IR or no 4IR.

increases their chances of becoming successful entrepreneurs. However, increased earnings, job performance, entrepreneurship, good health and well-being, and other outcomes of human capital are largely dependent on other numerous interrelated and somewhat complex factors. Some of these factors are contextual. Others are not and depend on the individual. Productivity and job performance, for example, also depend on the kind of work and other economic, social, and cultural systems in place in organizations and societies.

Collective Organizational Outcomes: Units, Firms, and Formal Organizations (Benefits to Formal Organizations)

Collective outcomes refer to human capital outcomes at the collective level such as units, firms or organizations, societies, and countries. This section focuses on benefits and outcomes to organizations, while the next section presents the outcomes and benefits to societies and countries.

Several studies have linked human capital to improved firm performance. In fact, understanding the link between human capital and firm performance forms the backbone interest of some academic disciplines. A core research interest of human resource management (HRM) and strategic human resource management (SHRM) is the effective management of people and their human capital for superior firm performance. Strategic human capital (SHC), one of the main groups within strategic management, also seeks to understand how firms utilize human capital for competitive advantage and superior performance. Crook et al.'s (2011) meta-analysis provides convincing evidence that human capital makes significant positive contributions to firm performance, especially when human capital is firm specific. In other words, the human capital developed within an organization is generally more productive within the context of that organization. It is pertinent to note that organizational or firm performance does not always equate or refer to financial performance and its diverse

measures such as profitability, return on investment, sales growth, and cash flows. While financial performance is an important part of organizational performance, other forms of performance such as innovation or ability to innovate processes, ideas, or products, productivity, and customer satisfaction are also indicators of firm performance, and all have been linked as outcomes of human capital (Crook et al., 2011). Human capital also boosts team performance within the organization (Hessels et al., 2020).

Conclusively, human capital, as firm resources either at the individual or more collective levels within the firm, contributes to firm performance and can help business firms to create competitive advantage. In fact, the most sustainable competitive advantage is the one that is based on people and their human capital (Bartlett & Ghoshal, 2002; Pfeffer, 1994). Effective management of people and their human capital lays the foundation for a truly sustainable competitive advantage (Pfeffer, 2005). All other business strategies will eventually be overridden with time as other businesses catch up if the strategies are not ultimately built on people and their individual and the firm's collective human capital. Indeed, no organization can exist without people and their human capital.

For organizations to optimize their human capital, they require good and effective human resource (HR) practices, policies, and systems. HR practices are activities directed toward the management of human resources, "the pool of human capital under the firm's control in a direct employment relationship" (Wright et al., 1994: p. 304). While HR practices are the actual human resource practices, HR policies "represent an organization's stated intentions about HR practices that should be implemented" (Boon et al., 2019: p. 2501). HR practices and policies are best formalized by having them expressly written rather than being mere oral traditions that guide the management of human resources or people within an organization. HR systems refer to the whole—the HR practices and policies—that guide the employment relationship between the organization and their employees. Certain groups of HR practices, regarded as high-performance work practices,[3] have been found to be particularly adept at increasing organizational performance (Boon et al., 2019; Combs et al., 2006).

The universality and contingency approaches to HR practices are two different approaches to the effectiveness and use of HR practices. While the main thrust of the contingency approach is that there has to be a strategic fit between HR practices and business strategy, the "universalist" approach adopts the best practices approach and emphasizes fit among HR practices as bundled into an HR system rather than fit with a specific business strategy (Delery & Doty, 1996). The implicit assumption in the "universalist" approach is that HR practices regarded as best practices can be applied in any organization. However, generally, HR practices need to be adapted to fit the purpose and strategy of the organization in order to result in positive organizational performance (Skaggs & Youndt, 2004; Youndt et al., 1996). The general consensus that has emerged on HR practices is that focus should be on HR systems rather than on individual HR practices (Boon et al., 2019). Overall, organizations with HR practices and policies that interrelatedly fit into a holistic whole have better HR systems than those that do not. Indeed, HR practices that do not form an interrelated whole can hardly be regarded as an HR system. An HR system includes practices and policies in HR functional areas such as recruitment and staffing, compensation and benefits, training and development, performance appraisal and management, and job definition and design.

There are three main mechanisms through which an effective HR system increases organizational performance. The first is by increasing the pool of human capital. This can be achieved through right recruitment and training and effective human capital development. The second is by enabling social exchanges between people, whom of course are the main embodiments of human capital. An effective means of achieving

this is the manner in which tasks are designed and the organization is structured. The third mechanism is by ensuring that employees are motivated enough to apply their human capital to productive work for the utmost benefit of the organization (Takeuchi et al., 2007). Providing adequate extrinsic and intrinsic motivation is one effective means of attaining this. The HR function in any organization is by implication, the management of the human capital of employees.

Organizations can also be conceptualized as having unit levels of capital—knowledge, skills, abilities, and other characteristics (KSAOs)—that enable the unit as a whole to be productive or particularly adept in specific areas. In this vein, organizational outcomes can also be said to be the results or outcomes of such unit-level human capital or capabilities. However, unit-level capabilities usually have their bedrock in the human capital—the KSAOs—of individuals. Moreover, the administration and orchestration of unit-level capabilities are conducted by individuals who undoubtedly are enabled by their human capital. Irrespective of the perspective through which the human capital in organizations is conceptualized, whether at the individual or collective levels, the cumulative benefits to organizations are enormous and reflected directly and indirectly in various positive organizational outcomes.

Collective Organizational Outcomes: Societies and Countries (Benefits to Societies and Countries)

Beyond the formal organizational level, human capital outcomes are also reflected at higher levels of organization of people. At societal and country levels, the impacts of human capital range from economic and health to other social outcomes. One of the best aggregations of people and one which generally provides ready forms of some measurements and which is also generally available is the country level. From a vast body of literature, human capital is one of the main drivers of economic growth of countries. Various empirical models especially in endogenous growth models in economics show that human

---

[3] High-performance work systems are sometimes referred to as high-involvement work practices or high-commitment work practices, though with slight variations in their meanings and perspectives.

capital contributes to the growth of gross national income (GNI) or gross domestic product (GDP) of countries. As far back as Solow's (1957) study, studies have demonstrated that economic growth is not the result of the accumulation of physical capital alone (Goldin, 2016). In fact, a huge proportion of the residual measuring the growth of countries could not be accounted for until human capital was included in the models (Goldin, 2016; Goldin & Katz, 2008; Mankiw et al., 1992). While there have been some arguments on the exact routes or mechanisms through which human capital impacts economic growth, the evidence is conclusive that human capital has a positive impact on economic growth and development (Acemoglu & Autor, 2012; Barro, 1991; Goldin, 2016; Goldin & Katz, 2008; Wilson & Briscoe, 2004). Through different econometric models, economic growth has been linked to human capital. A country's level of human capital, for example, is the driving force behind the success of new technological inventions and the absorption or adaptation of new technologies that fuel economic growth (Dao & Khuc, 2023).

However, it is pertinent to note that economic growth and development does not depend just on the levels (quantity and quality) of human capital. Other factors also come into play, and how human capital is deployed or utilized is also vital. It is possible to have lots of human capital both in terms of quantity and quality and not have optimal utilization of it because of the lack of enabling institutions, for example. Just as individuals and firms may not make utmost utilization of their human capital, countries are not exempted from falling into the same trap.

Measures of human capital in many extant studies linking human capital and economic growth have been quite diverse. This may help account for some of the arguments on the mechanisms through which human capital impacts economic growth and development. Some of the measures that have been used as proxy indicators of human capital of countries include share of education expenditure in GDP; years of primary, secondary, and tertiary schooling; share of university graduates in the workforce; and quality of public schools (Abraham & Mallatt, 2022). The

most common indicator used to measure human capital of countries is average years of schooling of a country's population and enrollment rates at different education levels (Abraham & Mallatt, 2022; Demirgüç-Kunt & Torre, 2022). These have been used largely because they are the most easily and readily available in most countries. When one considers that some components of human capital have hardly been reflected in the human capital studies linking human capital to economic growth, then one realizes that the impact of human capital on growth and development may have been grossly underestimated. Experience and training gained in the workplace, for examples, have not been included in most studies of human capital and economic growth and development. Knowledge, skills, and abilities gained from work and other experience can hardly be fully measured. Moreover, the effects of human capital on the development and advancement of technology are hardly accommodated in econometric models linking human capital to economic growth and development.

The impact of human capital on economic growth is both direct and indirect. Human capital increases the productivity of the labor force. Undoubtedly, a formally educated and trained workforce is much more productive than a workforce that is neither educated nor trained. The increased earnings that accrue to the individual, discussed in the last section, are a reflection of the increased productivity brought about by education and training, major components of human capital. Countries with quality human capital are also more competitive because of the innovation and diffusion of technology that human capital enables. An increase in human capital is likely to induce innovation and increase the number of innovative entrepreneurs as educated people are in the best position to invent, be entrepreneurial, be innovative, and have the capability to make use of advanced technologies (Diebolt & Hippe, 2019; Goldin & Katz, 2008). As an input in the research and development process, human capital increases the innovative and absorptive capacity of economies (Bye & Fæhn, 2022; Cohen & Levinthal, 1989). To borrow the words of Cohen and Levinthal (1989), a country's "ability to identify, assimilate, and exploit knowledge" (p. 569), or what they refer to as absorptive or learning capacity, is directly and indirectly related to its stock of relevant human capital.

Human capital also facilitates some positive outcomes and ensures that the direct positive outcomes of the factors are derived. Gains to trade and foreign direct investment (FDI), for examples, accrue more to countries with the requisite human capital than to those without (Aziegbe-Esho & Verhoef, 2023; Ahumada & Churata, 2019; Teixeira & Queirós, 2016). Many countries work toward attracting FDI, but some minimum levels of human capital are required for FDI to have a positive impact in host countries (Borensztein et al., 1998; Filippaios et al., 2019). When it comes to the ability of human capital to attract FDI or derive benefits from FDI and international trade, it has been shown that what matters is not necessarily the quantity of human capital in a country but the quality. Indeed, generally, when it comes to the impact of human capital on the economic growth and general well-being of societies, although quantity—the number of people with human capital—is important, quality is far more important. Countries with high levels of human capital generally have longevity, reflected in higher life expectancy rates, reduced infant mortality rates, and generally more economic development (Hanushek & Woessmann, 2008; Pelinescu, 2015; Wilson & Briscoe, 2004). Poverty is also generally lower in countries with high human capital levels. Summarily, human capital has a positive correlation with many positive societal outcomes and a negative correlation with negative societal outcomes.

In the competitive advantage of nations, and in describing the determinants of the prosperity of countries, Porter asserts that skilled human resources and scientific knowledge base of a country are the most important factors of production (Porter, 1990, 2000). In total agreement with this assertion, both of these factors—skilled human resources and scientific knowledge base—are predominantly human capital. A country's competitiveness is largely dependent on its ability to create, acquire, accumulate, and deploy human capital for productivity. The use to which all other resources and capital, including natural resources, can be put to is determined by the quantity and quality of human capital. Countries may be well endowed with diverse natural resources, but unless they have the capacity and ability to utilize the natural resources and convert them into productive use with their human capital, such diverse endowments amount to little in terms of benefits, especially to the overall benefits of the people. This argument is beyond the natural resource curse hypothesis of a negative correlation between natural resource abundance and economic growth (Sachs & Warner, 1995, 1997; Saeed, 2021). Indeed, there is nothing inherent in the abundance of natural resources that necessarily leads to poor economic performance (Mikesell, 1997). The main argument offered here, and one that perhaps lends some thesis to the natural resource curse, is that an abundance of natural resources without the requisite human capital to utilize and exploit the natural resources for the common good of the group or society, even with the best intentions of government and public policy, is a recipe for poor economic performance.

Finally, there are other positive externalities to having a population with human capital. Generally, when there is a high level of human capital in a country, the electorate are usually more informed and crime rate can be lower (Abraham & Mallatt, 2022). There are evidences that societies with higher levels of human capital have greater social cohesion and lower infant mortality and participate more in the electoral process in democracies (Kavanagh & Doyle, 2006).

The Dynamics of Human Capital Outcomes

All outcomes of human capital, whether direct or indirect, depend on some individual and group dynamics. The amount of increased earnings and wages, for instance, depend on factors such as the specific nature of human capital, the demand and supply in human capital markets,[4] the bargaining power of individuals, and of course, other personal and contextual factors surrounding the individual. These factors are also interrelated. A person with a set of skills that are in high demand by companies and organizations, for instance, will most likely be able to bargain and earn higher wages and other conditions of employees' service than one with a set of outdated skills not readily required by companies. A situation where the supply for human capital exceeds that of the demand for it will most certainly affect the private returns in the form of wages and salaries that can accrue to individuals.

In addition to the complexity in respect to returns to individuals on their human capital, the dynamics of human capital outcomes at group levels are a bit more complex. As shown above in this chapter, there are also several group dynamics that determine the benefits that groups can obtain from having individuals with human capital. Merely having persons with human capital will not automatically translate to positive group outcomes, no matter the quality or nature of the human capital capacities embodied in the individuals. Therefore, at the group level, deriving benefits from human capital requires deliberate management of the stock of human stock and of the individuals. Organizations will not become more profitable or improve their various performance indexes instantaneously just by employing persons with human capital. The same applies to countries. Countries do not grow and become developed just by having persons with human capital. Systems and processes that ensure that human capital is put to productive use for the good of both the individual and the group have to be put in place and enforced in order for group

outcomes of human capital to find manifestations.

One of the essential ingredients for unleashing the potential of human capital in groups is good institutions. One of the most popular definition of institutions is that given by North (1990) in which he defines institutions as "the rules of the game of a society" (p. 3). At the group level, the rules of the game of any group affect the human capital outcomes of the group both for the persons within the group and for the group as a whole. The discussion on HR practices and systems earlier in this chapter on collective outcomes at the unit, firm, and organizational levels aptly portrays this. HR practices are a form of the institutions in a formal organization. HR practices are a form of "the humanly devised constraints that structure human interaction" (North, 1990, p. 3) in formal organizations. Similar to formal organizations, societies also have formal and informal institutions. However, the complexity of societies "make the rules of the game" much more complex than that of formal organizations. Consequently, "the play of the game" (Filippaios et al., 2019; Williamson, 1998) interacts with "the rules of the game" (North, 1990) and various other factors including informal institutions to determine human capital outcomes in societies in complex manners.

The Micro–Macro Paradox and Human Capital Outcomes

The role of institutions in human capital outcomes in groups and organizations such as countries highlights the often discussed "micro–macro paradox" of education and human capital. There is little doubt that education and human capital are beneficial to individuals (e.g., Psachararopoulos, 1994; Psachararopoulos & Patrinos, 2004). However, the effect of more education, specifically more formal schooling, to economic growth is a bit more controversial. While several studies have found and established a positive link (e.g., Barro, 1991; Wilson & Briscoe, 2004), others have found no significant effects, and some have even found negative

---

[4]The market for human capital and the labour market are similar but not exactly synonyms. The market for human capital refers to the content, processes, and institutions that govern the acquisition and development of KSAOs that qualify as human capital (Esho & Verhoef, 2020). The labour market can therefore be regarded as a subsection of the market for human capital. Esho and Verhoef (2020) provide more ample explanations on the market for human capital.

effects (Pritchett, 2001; Schündeln & Playforth, 2014). This represents the "micro–macro paradox" of human capital outcomes in which aggregate outcomes are far less than what is expected given the positive outcomes at the individual level. In other words, the private returns to education that is seen in individuals do not correspond to the expected aggregate outcomes at the country level—the effect of more schooling on economic growth is far less than expectations. The evidence of countries such as India and Venezuela give further empirical credence to this paradox (Schündeln & Playforth, 2014). Pritchett (2001) puts forward three non-mutually exclusive possibilities for this paradox. First, the human capital created may have been deployed into unproductive activities. Second, the supply of human capital becomes more than the demand. Third, perhaps no human capital was created through formal schooling. Schooling does not always equate to learning (Prichett, 2013). Any one or more of these reasons can account for the great private returns that seem to elude public or group returns to education and human capital. One or more of these reasons may also account for the varying degrees of private returns to human capital.

This "micro–macro paradox" highlights the difference between private and public returns to education and more broadly human capital. The returns to human capital are both private and public. Along each of these two dimensions are varying degrees of returns. For simplicity, a two-by-two matrix may help to present the interactions between these two types of returns. Each type of returns, private and public, can either be low or high. Ideally, individuals and groups desire the highest form of returns, quadrants 1 and 3 for individuals and 1 and 2 for groups (see Fig. 4.1).

People may deploy their human capital to maximize their private returns without being mindful of whether such deployment yields any public return to the collective group to which they belong or in which they make such deployment. In fact, it is doubtful if individuals with human capital actually think of the public returns.

This does not preclude educated people with lots of human capital finding employment in sectors where the private returns to their education are low, quadrants 2 and 4 in Fig. 4.1 (see Schündeln & Playforth, 2014). Consequently, depending on the nature of returns to human capital, outcomes may not be obvious at the aggregate level such as in economic growth. All things being equal, while individuals will seek to maximize their private returns (quadrant 3 or 1), groups will seek to maximize the aggregate returns to the group (quadrant 1 or 2). In the final analysis, the onus lies with the individual and

Fig. 4.1 The human capital returns (HCR) matrix

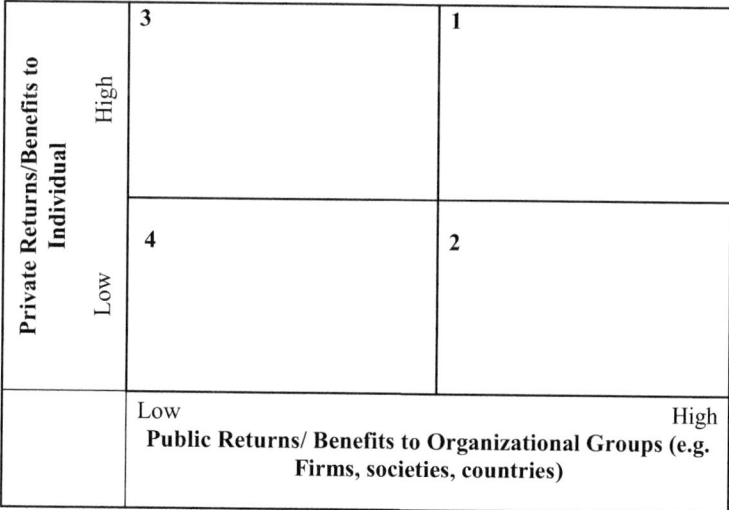

specific groups to ensure that they derive benefits from their stock of personal and group human capital. This analysis may be further complicated by the type of investments, whether private or public investments, made in acquiring and developing the human capital. Where special public investments in education and human capital of individuals are made, guidelines and policies that ensure and facilitate public returns to such investments can be means of ensuring that there are enough group returns to such investments.

Empirical studies on the effect of human capital investments on economic growth, and industrial development, in African countries currently present very mixed evidence. While some more recent studies have found positive effects (e.g., Ewane & Ewane, 2024; Ofori et al., 2024; Sajoh, 2021; Wirajing et al., 2023), others, mostly earlier studies, found negative or inclusive evidence (e.g., Appleton & Teal, 2002; Brunschwig et al., 1998; Karambakuwa et al., 2020). These studies do not specifically investigate the micro–macro paradox of human capital outcomes. However, they are indications that the positive effects of human capital at aggregate or macro levels are rather complex and may differ from returns and benefits that individuals derive.

Conclusion

To conclude, the individual and collective returns and benefits discussed in this chapter implicitly assume that the individual, group, or collective has the right quantity, level, type, and quality of human capital. Otherwise, the discussed benefits may never materialize or may take longer to emerge and become visibly felt within and outside the group. The benefits and utility of human capital also depend on its nature and its different components and dimensions. Indeed, the benefits and outcomes of human capital depend on multiple complex dynamics. Summarily, all human capital are not equal. Consequently, the benefits and impact depend on numerous factors that relate to the human capital and on the context in which it has both been developed and is being deployed and utilized for productive activities.

References

Abraham, & Mallatt. (2022). Measuring human capital. *Journal of Economic Perspectives, 36*(3), 103–130.

Acemoglu, D., & Autor, D. (2012). What does human capital do? A review of Goldin and Katz's the race between education and technology. *Journal of Economic Literature, 50*(2), 426–463.

Ahumada, V. M. C., & Churata, R. I. J. L. (2019). The effects of nafta on economic growth. *Investigacion Economica, 78*(308), 63–88.

Appleton, S., & Teal, F. (2002). *Human capital and economic development*. Working paper series 173, working paper 39, African Development Bank.

Aziegbe-Esho, E., & Verhoef, G. (2023). Reaping the benefits of African Continental Free Trade Agreement (AfCFTA): The role of human capital development. *Africa Review, 15*, 1–23.

Barro, R. J. (1991). Economic growth in a cross section of countries. *The Quarterly Journal of Economics, 106*(2), 407–443.

Bartlett, C. A., & Ghoshal, S. (2002). Building competitive advantage through people. *MIT Sloan Management Review, 43*(2), 34–41.

Becker, G. S. (1964). *Human capital: A theoretical and empirical analysis, with special reference to education*. University of Chicago Press.

Becker, G. S. (2007). Health as human capital: Synthesis and extensions. *Oxford Economic Papers, 59*, 379–410.

Bloom, D., & Canning, D. (2003). Health as human capital and its impact on economic performance. *The Geneva Papers on Risk and Insurance, 28*(2), 304–315.

Boon, C., Hartog, D. N. D., & Lepak, D. P. (2019). A systematic review of human resource management systems and their measurement. *Journal of Management, 45*(6), 2498–2537.

Borensztein, E. J., De Gregorio, J., & Lee, J.-W. (1998). How does foreign direct investment affect economic growth? *Journal of International Economics, 45*(1), 115–135.

Brunschwig, S., Sacerdoti, E. & Tang, J. (1998). *The impact of human capital on growth: Evidence from West Africa*. IMF working paper WP/98/162, November 1998.

Bye, B., & Fæhn, T. (2022). The role of human capital in structural change and growth in an open economy: Innovative and absorptive capacity effects. *The World Economy, 45*, 1021–1049.

Cohen, W. M., & Levinthal, D. A. (1989). Innovation and learning: The two faces of R&D. *Economic Journal, 99*, 569–596.

Combs, J., Liu, Y., Hall, A., & Ketchen, D., Jr. (2006). How much do high-performance work practices matter? A meta-analysis of their effects on organizational performance. *Personnel Psychology, 59*, 501–528.

Crook, T. R., Todd, S. Y., Combs, J. G., Woehr, D. J., & Ketchen, D. J., Jr. (2011). Does human capital matter? A meta-analysis of the relationship between human capital and firm performance. *Journal of Applied Psychology, 96*(3), 443–456.

Dao, T. B. T., & Khuc, V. Q. (2023). The impact of openness on human capital: A study of countries by the level of development. *Economies, 11*(7), 175. https://doi.org/10.3390/economies11070175

Delery, J. E., & Doty, D. H. (1996). Modes of theorizing in strategic human resource management: Tests of universalistic, contingency, and configurational performance predictions. *Academy of Management Journal, 39*(4), 802–835.

Demirgüç-Kunt, A., & Torre, I. (2022). Measuring human capital in middle income countries. *Journal of Comparative Economics, 50*(4), 1036–1067. https://doi.org/10.1016/j.jce.2022.05.007

Diebolt, C., & Hippe, R. (2019). The long-run impact of human capital on innovation and economic development in the regions of Europe. *Applied Economics, 51*(5), 542–563. https://doi.org/10.1080/00036846.2018.1495820

Dimov, D. (2017). Towards a qualitative understanding of human capital in entrepreneurship research. *International Journal of Entrepreneurial Behaviour & Research, 23*(2), 210–227. https://doi.org/10.1108/IJEBR-01-2016-0016

Esho, E., & Verhoef, G. (2020). A holistic model of human capital for value creation and superior firm performance: The strategic factor market model. *Cogent Business & Management, 7*(1). https://doi.org/10.1080/23311975.2020.1728998

Ewane, E. B., & Ewane, E. I. (2024). Human capital development and industrial sector growth in Sub Saharan African countries. An augmented pooled mean group estimator. *Arab Economic and Business Journal, 16*(2), Article 2. https://doi.org/10.38039/2214-4625.1047

Filippaios, F., Annan-Diab, F., Hermidas, A., & Theodoraki, C. (2019). Political governance, civil liberties, and human capital: Evaluating their effect on foreign direct investment in emerging and developing countries. *Journal of International Business Studies, 50*, 1103–1129.

Filmer, D., Gatti, R., Rogers, H., Spatafora, N., & Emrullahu, D. (2021). *Education and health for inclusiveness.* IMF working paper WP/21/60.

Goldin, C. (2016). Human capital. In C. Diebolt & M. Haupert (Eds.), *Handbook of cliometrics.* Springer Verlag.

Goldin, C., & Katz, L. F. (2008). *The race between education and technology.* The Belknap Press of Harvard University.

Hanushek, E. A., & Woessmann, L. (2008). The role of cognitive skills in economic development. *Journal of Economic Literature, 46*(3), 607–668.

Hessels, J., Rietveld, C. A., Thurik, A. R., & Van der Zwan, P. (2020). The higher returns to formal education for entrepreneurs versus employees in Australia. *Journal of Business Venturing Insights, 13*, e00148.

Karambakuwa, R. T., Ncwadi, R., & Phiri, A. (2020). The human capital-economic growth nexus in SSA countries: What can strengthen the relationship? *International Journal of Social Economics,* 47(9), 1143–1159. https://doi.org/10.1108/IJSE-08-2019-0515

Kavanagh, C., & Doyle, E. (2006). *Human capital and productivity in the Irish context* (pp. 1–95). Expert Group on Future Skills Need.

Mankiw, G., Romer, D., & Weil, D. (1992). A contribution to the empirics of economic growth. *Quarterly Journal of Economics, 107*, 407–438.

Martin, B. C., McNally, J. J., & Kay, M. J. (2013). Examining the formation of human capital in entrepreneurship: A meta-analysis of entrepreneurship education outcomes. *Journal of Business Venturing, 28*(2), 211–224.

Mikesell, R. F. (1997). Explaining the resource curse, with special reference to mineral-exporting countries. *Resources Policy, 23*(4), 191–199.

Mincer, J. (1974). *Schooling, experience, and earnings.* New York, National Bureau of Economic Research.

Moll, I. (2021). The myth of the fourth industrial revolution. *Theoria, 68*(167), 1–38.

North, D. (1990). *Institutions, institutional change and economic performance.* Cambridge University Press.

Ofori, P. E., Kuuwill, A., & Quaye, B. (2024). Effect of human capital development and institutional quality on inclusive growth in African countries. *Cogent Economics & Finance, 12*(1). https://doi.org/10.1080/23322039.2024.2357155

Pelinescu, E. (2015). The impact of human capital on economic growth. *Procedia Economics and Finance, 22*, 184–190.

Pfeffer, J. (1994). Competitive advantage through people. *California Management Review, 36*(2), 9–28.

Pfeffer, J. (2005). Producing sustainable competitive advantage through the effective management of people. *Academy of Management Executive, 19*(4), 95–106.

Porter, M. (1990). The competitive advantage of nations. *Harvard Business Review*, March–April, 73–91. Available at https://economie.ens.psl.eu/IMG/pdf/porter_1990_-_the_competitive_advantage_of_nations.pdf

Porter, M. E. (2000). Attitudes, values, beliefs and the microeconomics of prosperity. In L. E. Harrison & S. P. Huntington (Eds.), *Culture matters: How values shape human progress.* Basic Books.

Prichett, L. (2013). *The rebirth of education: Schooling ain't learning.* CGD Books, Centre for Global Education.

Pritchett, L. (2001). Where has all the education gone? *The World Bank Economic Review, 15*(3), 367–391.

Psachararoupoulos, G. (1994). Returns to investment in education: A global update. *World Development, 22*(9), 1325–1343.

Psachararoupoulos, G., & Patrinos, H. A. (2004). Returns to investment in education: A further update. *Education Economics, 12*, 111–134.

Rifkin, J. (2012). The third industrial revolution: How the internet, green electricity, and 3-D printing are ushering in a sustainable era of distributed capitalism. *The World Financial Review.*

Rifkin, J. (2016). *The 2016 World Economic Forum misfires with its fourth industrial revolution theme*. Available at https://www.industryweek.com/technology-and-iiot/information-technology/article/21967057/the-2016-world-economic-forum-misfires-with-its-fourth-industrial-revolution-theme

Sachs, J. D., & Warner, A. M. (1995). *Natural resource abundance and economic growth*. NBER working paper series.

Sachs, J. D., & Warner, A. M. (1997). Sources of slow growth in African economies. *Journal of African Economies, 6*(3), 335–376. https://doi.org/10.1093/oxfordjournals.jae.a020932

Saeed, K. A. (2021). Revisiting the natural resource curse: A cross-country growth study. *Cogent Economics & Finance, 9*(1), 2000555. https://doi.org/10.1080/23322039.2021.2000555

Sajoh, A. A. (2021). The effect of human capital on economic growth in some Sub Sahara African countries (SSA). *American Journal of Economics, 5*(1), 1–24. https://doi.org/10.47672/aje.884

Schultz, T. W. (1961). Investment in human capital. *The American Economic Review, 51*(1), 1–17.

Schündeln, M., & Playforth, J. (2014). Private versus social returns to human capital: Education and economic growth in India. *European Economic Review, 66*, 266–283. https://doi.org/10.1016/j.euroecorev.2013.08.011

Skaggs, B. C., & Youndt, M. (2004). Strategic positioning, human capital, and performance in service organizations: A customer interaction approach. *Strategic Management Journal, 25*, 85–99.

Solow, R. (1957). Technical change and the aggregate production function. *Review of Economics and Statistics, 39*, 312–320.

Takeuchi, R., Lepak, D. P., Wang, H., & Takeuchi, K. (2007). An empirical examination of the mechanisms mediating between high-performance work systems and the performance of Japanese Organizations. *Journal of Applied Psychology, 92*(4), 1069–1083.

Teixeira, P. N. (2014). Gary Becker's early work on human capital – Collaborations and distinctiveness. *IZA Journal of Labor Economics, 3*, 12. https://doi.org/10.1186/s40172-014-0012-2

Teixeira, A. A. C., & Queirós, A. S. S. (2016). Economic growth, human capital and structural change: A dynamic panel data analysis. *Research Policy, 45*(8), 1636–1648.

Unger, J. M., Rauch, A., Frese, M., & Rosenbusch, N. (2011). Human capital and entrepreneurial success: A meta-analytical review. *Journal of Business Venturing, 26*, 341–358.

Walsh, J. R. (1935). Capital concept applied to man. *The Quarterly Journal of Economics, 49*(2), 255–285.

Williamson, O. E. (1998). The institutions of governance. *The American Economic Review, 88*(2), 75–79.

Wilson, R. A., & Briscoe, G. (2004). The impact of human capital on economic growth: A review. In P. Descy & M. Tessaring (Eds.), *Impact of education and training, third report on vocational training research in Europe: Background report*. Office for Official Publications of the European Communities. (Cedefop reference series, 54).

Wirajing, M. A. K., Nchofoung, T. N., & Etape, F. M. (2023). Revisiting the human capital-economic growth nexus in Africa. *SN Business & Economics, 3*, 115. https://doi.org/10.1007/s43546-023-00494-5

Wright, P. M., McMahan, G. C., & McWilliams, A. (1994). Human resources and sustained competitive advantage: A resource-based perspective. *International Journal of Human Resource Management, 5*(2), 301–326.

Youndt, M. A., Snell, S. A., Dean, J. W., & Lepak, D. P. (1996). Human resource management, manufacturing strategy and firm performance. *Academy of Management Journal, 39*, 836–866.

Zacharakis, A. L., & Meyer, G. D. (2000). The potential of actuarial decision models: Can they improve the venture capital investment decision? *Journal of Business Venturing, 15*(4), 323–346.

Part II

Case Studies of National Outcomes of Strategic Human Capital Development

Global Human Capital Indexes and Introduction to Part 2

5

Abstract

This chapter is focused on some global human capital indexes and the different indicators that have been used for its measure by the indexes. Indicators range from school enrolment rates, years of schooling, survival rates, mortality rates, and unemployment rates. These indicators relate to knowledge and education and health dimensions of human capital. Clearly, it is quite difficult for any one indicator to fully capture a multidimensional construct such as human capital. However, the indexes have attempted to reflect as much as possible the prevailing state of countries' human capital. The indexes, using different data and methodologies, have ranked each country to an ideal rather than mere comparisons among countries. All indexes have their utility. However, while some may be more suited for academic research and analysis, others may be more suitable for business decision-making and guidance for policy formulation. Although African countries generally rank poorly in all the indexes, there is still an opportunity to realize the potential in Africa's greatest resource—the youthful population.

Keywords

Global index · Human capital · HCI · HDI · GTCI

Introduction

Human capital is a multicomponent, multidimensional construct. It is also composite and cannot be directly measured. Its existence can only be observed by indicators. Consequently, human capital cannot truly be measured precisely. Any measurement of human capital quite often depends on the purpose for which it is being measured. It also sometimes depends on who is doing the measuring. Economists, for example, focus primarily on education measured at the country level and as a result measure human capital with various forms of formal schooling metrics such as enrollment rates and school completion rates at the three main tiers of education—primary, secondary, and tertiary. Strategy and human resource management scholars rely on proxies of education and work experience such as tenure at a job role, business organization, industry, or occupation. The type of measurement and levels at which human capital is measured depends on differing interests. Quite clearly, for the purpose of assessing national outcomes, the economists' approach is the most preferred. This invariably means that large portions of human capital of a country such as the work experience of the population are left unmeasured. Some economics studies that have measured human capital using work experience have shown that the contribution of physical and human capital to national income increases from 40% to 60% when human capital from work experience is taken into account (Lagakos et al., 2014). Still, work experience is scarcely used in measuring human capital of countries. A foremost merit of the economists' approach of using measures of formal education is that such measures are usually indicative of the potentials of a country rather than the mere valuation of the present levels of human capital. Consequently, a number of academic studies measuring the stock of human capital for countries primarily through education have been published (e.g., Barro & Lee, 1993, 2001, 2013; Cohen & Soto, 2007; Hanushek & Woessmann, 2012; Lagakos et al., 2014). While most have concentrated on using quantity indicators of measurement, others have incorporated work experience (e.g., Lagakos et al., 2014) and taken cognizance of quality of formal education (e.g., Hanushek & Woessmann, 2012).

Global Indexes of Human Capital

Research centers in some universities have certain measurements for countries' human capital. One of such indexes is that of the Institute for Health Metrics and Evaluation (IHME) at the University of Washington, United States (USA) (Lim et al., 2018). Lim and colleagues developed a period measure of expected human capital. They rank countries according to the estimated number of years lived from 20 to 64 years, and the measure has been adjusted for educational attainment, learning, and functional health. So, a country like Finland that tops the rankings for 1990 and 2016, years presented on the index, and has expected human capital years of 28 years in 2016 is much better ranked than another country that has lower number of years, say 20 years, for example. Consequently, a person born in Finland is expected to have effective human capital years of 28 years on the average. Another measure is that of the Penn World Table (PWT) which was originally developed at the University of Pennsylvania (Summers & Heston, 1988) and now hosted at *the* University of California, Davis, *and the* University of Groningen, the Netherlands. Though not primarily a database for human capi-

tal but for national inputs, outputs, and productivity levels, some versions contain data on human capital for the different countries included in the database. From PWT version 8, data on human capital calculated based on average years of schooling and returns to education has been included (Feenstra et al., 2015). So, from the data, countries with higher average years of schooling are assumed to have more human capital.

Insead Business School, France, in collaboration with Portulans Institute, USA, and Human Capital Leadership Institute, Singapore, have since 2013 published an annual index, the Global Talent Competitiveness Index (GTCI). GTCI is a composite index that measures the "set of policies and practices that enable a country to develop, attract, and empower the human capital that contributes to productivity and prosperity" (GTCI, 2023: G 14). GTCI is, therefore, not just a measure of countries' stock of human capital. It also measures the potential for effective deployment of the stock of human capital. The index is tilted more toward measuring country's "talent competitiveness" but can also be said to be a reflection of the human capital available in the countries. It can also be regarded as a measure that somewhat reflects the quality of human capital available in the countries included in the index. The GTCI is a human capital index that also reflects human capital deployment and productivity levels of human capital in a country.

These publications all aim at quantifying the level of human capital globally in countries. While each of the aforementioned indexes has its utility, the human capital data of PWT and IHME may be more suited for academic research and analysis. GTCI seems to be more suitable for competitive decision-making for global companies and businesses as it comprises direct measures of human capital as well as what can be regarded as human capital facilitators. Included in GTCI as human capital facilitators are the regulatory, market, business, and "labor" landscape of each country as well as indicators of degree of openness of each country. Together, the direct and indirect measures enhance the ability of countries to attract, grow, retain, and enable people to effectively acquire, develop, and deploy their human capital. Similar to the GTCI are other global indexes computed by nonacademic global agencies such as the World Bank and United Nations. Three of these indexes are presented in greater detail below.

Other Global Human Capital Indexes There are three main global indexes computed by global agencies aimed at tracking the human capital development of countries: World Bank's Human Capital Index (HCI), the World Economic Forum (WEF)'s Global Human Capital Index (GHCI), and United Nations' Human Development Index (HDI).

Human Development Index (HDI) HDI is a composite measure of people's capabilities introduced in 1990 by the United Nations (UN). Of the three global indexes by global agencies, it is the oldest and was created to emphasize that true development of any country results from people and their capabilities. "People are the real wealth of nations" (UN, 2022). So, while the importance of economic growth cannot be overemphasized, HDI emphasizes that true development should be measured by how developed people actually are or people's human capital. Essentially, without human capital development, there is no real development. HDI can be regarded as much more than a composite measure; it is also actually a composite index made up of education component measures and measures of longevity and income. Longevity is measured using life expectancy at birth, an indicator of good health. To measure education, an indicator of knowledge, the core component of human capital, HDI uses expected years of schooling of children of school ages and mean years of schooling of adults of 25 years of age and above. An indicator of the standard of living individuals have in their countries is the third component of HDI. This is measured by the gross national income per capita based on each country's purchasing power parity in US dollars. HDI score ranges from the "conceivable" lowest score of zero (0) to the possible highest score of one (1), for over 190 countries

and territories. Data sources for HDI are diverse and include data and calculations from Barro and Lee, foremost academic experts of human capital computations.

Global Human Capital Index (GHCI) GHCI was first published in 2013 by the World Economic Forum (WEF) to measure the human capital of countries. Last published in 2017, the index ranked about 130 countries on a scale of 0, worst, to 100, best. In the years of publication, the index was based on certain indicators of education, health, and work experience. In its first edition, the index measured human capital by focusing on four pillars: education, health and wellness, workforce and employment, and the enabling environment in each country. The last independent report of the index in 2017 measured countries' human capital according to four equally weighted elements: capacity, deployment, development, and know-how (Fraumeni & Liu, 2021). Each of these four major areas are subindexes with different indicators as measurements. "The Capacity subindex quantifies the existing stock of education across generations; the deployment subindex covers skills application and accumulation of skills through work; the development subindex reflects current efforts to educate, skill and upskill the student body and the working age population; and the Know-how subindex captures the breadth and depth of specialized skills use at work" (GHCI, 2017: p. V11). The capacity element is akin to education in the UN's HDI but measured using the percentage of a country's population that has attained at least primary, secondary, and tertiary education. Basic literacy and numeracy levels of a country's population are also measured as part of the capacity element. The deployment element attempts to capture human capital derived from work experience, an element that can be said to be missing from the HDI. To capture human capital deployment, GHCI uses prevalent employment and unemployment rates of countries. The rate of underemployment and gender gap in participation in economics activities are also included in the deployment element. To capture the third element, development, net school enrolment rates in primary, secondary, and tertiary schools as well as quality of education, staff training, and skills diversity of countries were utilized. The final and fourth element is know-how. GHCI measured this subindex using the level of middle- and highly skilled workforce and economic complexity of a country.

Although there were changes to the indicators used for measuring human capital in the GHCI during the years of publication, the general approach was measuring different components of human capital across different age groups.[1] By evaluating the level of human capital across different age groups, WEF refers to this as a "life-course approach" as it took into cognizance that there is no age limit to human capital development and deployment—invariably learning, acquiring, and deploying human capital can happen throughout the course of an individual's life. Another merit of the GHCI is that it measures the stock of human capital in addition to potential. It also attempts to measure work experience, thus going beyond a focus on education which has been the main area of focus of academic measures. The index also went beyond measuring just quantity of education by incorporating some indicators of quality education. Although now discontinued as a separate report, a mini-version of the GHCI without any ranking is now incorporated into WEF's annual competitiveness report.

Human Capital Index (HCI) HCI was launched in 2018 by the World Bank to capture the health and education that a child born today is expected to achieve by the age of 18. The index takes cognizance of the risks of poor education and poor health prevailing in each country on the index (Yusuf, 2020). In other words, the index measures the potential productivity of the next generation of workers (Corral et al., 2021). HCI is one of the outputs of World Bank's Human Capital Project (HCP) that aims to accelerate global investments in people. The index comprises two major components of human capital: education and health. Over the years of the index, since 2018, it has measured education using years of learning and measured health with mortality rates. To measure quality of education, HCI includes a measure of harmonized test scores across countries. In 2019, the World Bank introduced the Socioeconomically Disaggregated Human Capital Index (SES-HCI) for 50 low and medium-income countries to enable better comparison across income groups within "similar" countries.

The three main global indexes enumerated above can be somewhat used to gauge both the effectiveness of national education policies and other policies that affect people, growth, and development. This undermines the importance of human capital as a major tool that can be leveraged for growth and development. Of the three global human capital indexes, the GHCI appears more comprehensive. It also aligns most with the theory of human capital as it attempts to measure the stock of a country's human capital in addition to its development and deployment by the individual and by countries. Unlike the other two indexes, the GHCI also incorporates human capital garnered from work experience. It can be argued that HDI's use of gross national income per capita to measure income can be regarded as an implicit measure of work experience. However, GHCI's measure of income appears to be more direct than that of HDI.

The weaknesses of each of the indexes notwithstanding (see Table 5.1 for a comparison of HDI, GHCI, and HCI), each has its utility and allows countries to have an idea of their level of human capital and gaps in their human capital development and deployment. The indexes are also useful for comparison across countries, and the actions and policies of countries in higher ranks can serve as learning points for lower ranked countries.

---

[1]Please refer to the GHCI reports for detailed information on this and other related information on the index. Also, refer to individual reports of all the indexes presented in this chapter for more details on each human capital index.

Table 5.1 A comparison of the three "nonacademic" global human capital indexes

Name of index	Human Development Index (HDI)	Global Human Capital Index (GHCI)	Human Capital Index (HCI)
Curating institution	United Nations	World Economic Forum (WEF)	World Bank
Dimensions of human capital	Education, health, and income	Education, health, and work experience	Education and health
Sample indicators employed	Education: Expected years of schooling, mean years of schooling Health: Life expectancy Income/standard of living: Gross national income per capita	Education: School enrollment rates, youth literacy rates, Internet access in schools Work experience: Employment, unemployment, and underemployment rates Health: Life expectancy	Education: Expected years of learning Health: Adult survival rates, rate of stunting for children 5 years and under
Measures	Relatively unchanged	Evolved	Relatively unchanged
First year of coverage	1990	2013	2018
Year of launch	1990	2013	2018
Number of countries ranked	>190	130	>157
Value range of index	0–1 (worst to best)	0–100 (worst to best)	0–1 (worst to best)
State of report	Active	Discontinued as an independent report	Active

Source: Author's compilation

Regional Indexes

The European Human Capital Index (EHCI) This is an index ranking of the human capital of European Union (EU) member countries. It was created by the Lisbon Council, the Brussels-based policy think tank, and Deutschland Denken, a Germany-based think tank. EHCI seeks to rank each EU's country according to their ability to nurture and develop their human capital in terms of their human capital endowment, human capital utilization, human capital productivity, and demographic outlook (Ederer, 2006; Ederer et al., 2007). EHCI was first launched in 2006, and a subsequent edition ranking of Central and Eastern EU countries was conducted in 2007. The first edition in 2006 ranked 13 EU countries. Notable about the EHCI is the recognition that the sources of human capital endowment go beyond formal schooling. The index takes cognizance of parental education and learning on the job as sources of human capital.

Europe and Central Asia Human Capital Index (ECA-HCI) This index, published in 2022, was developed by Demirgüç-Kunt and Torre (2022) to reflect some specific education and health conditions that are peculiar to middle-income countries in Asia and Europe. The index relies on World Bank's HCI, making adjustments for risk factors such as the prevalence of obesity, smoking, and heavy drinking in middle-income countries in these two regions.

Utility of Human Capital Indexes

Individuals are the real owners of their human capital, but its deployment and utilization depend on factors relating to both the individual and the organizational context (firm, society, other geographical contexts) in which the individual resides and deploys the human capital (Esho & Verhoef, 2020; Kim & Mahoney, 2007). The context in which people acquire human capital is a huge determining factor on the nature of human capital acquired and the context in which human capital deployed in work also affects its productivity and manner in which it is deployed by individuals and organizations. National frameworks and policies for education, health, and work (especially work related to business organizations) have a direct and invaluable impact on human capital acquisition, development, and deployment. Consequently, global human capital indexes may actually not accurately reflect the state of a country's human capital. It is in the light of this that regional indexes have been developed to more accurately reflect the peculiarities of some countries and regions. Having said this, countries' rankings on any of the indexes are not accidental. Indeed, there are strong correlations between the indexes. Reported independent calculations of correlations between HDI, GHCI, and HCI ranged from 0.85 to 0.95 (see Liu & Fraumeni, 2020; Abraham & Mallatt, 2022). This shows that a country that is ranked poorly in one index is most likely to also be poorly ranked in the other indexes. A cursory check on the three academic indexes also highlights the same fact. The top performing countries and the lowest ranked countries are nearly the same across all six indexes. Considering the wide variety of sources, components, measures, and methodologies used for these indexes, the general convergence of the indexes indicates that the scores and rankings are a good reflection of the state of human capital in the different countries.

Critiques and Criticisms of the Global Human Capital Indexes

The global indexes have been criticized by some. The World Bank's HCI, especially, has come under many criticisms, perhaps because of the highly publicized release of the index. Countries such as India notably rejected the index when it was first released in 2018, citing gaps in data and methodological weaknesses in the index and the need for better measures in this digital era (Business Standard, 2018; Economic Times, 2018; Shukla, 2018). Some of the arguments

against the index are related to extant criticisms of the notion of human capital itself and the resultant theory on its effects on income, economic growth, and other outcomes. They are, therefore, not too different from the criticisms that have been labeled against human capital discussed at the end of Chap. 2.

The construction of regional indexes by some countries and regions to make adjustments and allowances for peculiarities is also proof that global human capital indexes are indeed somewhat deficient. The indexes are rather too general and may not take adequate cognizance of other specific conditions that may be prevalent in countries and regions. Questions have also been raised on some of the measures adopted by the HCI (e.g., Liu, 2018; Liu & Steiner-Khamsi, 2020). For example, Liu and Steiner-Khamsi (2020), while not entirely against the HCI, argue that the use of international large-scale assessments, one of the measurements of education in the first edition, penalizes nonparticipating countries or countries that only partially participate in such large-scale assessments. Others relate to the multiplicity of global indexes for human capital (Edwards, 2018; Hunter & Shaffer, 2022) arguing that the indexes do not really solve any problem but rather are unnecessary duplication of efforts (Edwards, 2018). Still, others, while lauding the HCI, give concerns relating to its subsumption of health care to GDP growth, warning that this may lead to inequality in accessing health care and the financialization of health, a situation where financial actors, motives, markets, and institutions determine the types of health care that are available (Stein & Sridhar, 2019).

The different indexes do indeed have some methodological weaknesses, and some of these arise out of the multidimensional nature of human capital. However, as already mentioned above, the high correlation between most of the indexes, despite the diverse methodologies and data sources, somewhat lend credibility to the indexes. Moreover, the indexes are only analytical tools that reflect the overall state of human capital development and accumulation in a country. The indexes are meant to be tools of references for policy and decision-makers, and will be more useful as such, rather than as tools for mere comparison between and among countries. Therefore, countries will do well to use the indexes as gauges and tools for assessing their human capital conditions in terms of levels, quality, and dimensions of human capital development while bearing the limitations of the indexes in mind. Alternatively, countries may also construct their own human capital indexes that may better measure and reflect the state of their countries' human capital.

Despite the critiques and criticisms that have been leveled against the global indexes, they provide useful lenses through which countries can view their stock of human capital. The diverse measures used in the different indexes are indicators of the diverse dimensions, types, and components of human capital. This diversity also points to the diverse effects, benefits, and impacts that can potentially result from human capital. The indexes provide a means through which strategic planning and development on investments in people can be conducted and continuously evaluated to increase the chances of reaping the potential benefits of human capital investments. Finally, the validity of the global human capital indexes can also be seen through the lens of other global indexes, especially those that attempt to measure the wealth or prosperity and economic conditions of countries. Despite the huge diversities in the methodologies adopted, and sometimes in the purpose of the index, simple comparative analyses of the country rankings show that the same countries are almost always ranked at the top and bottom across most of the indexes.

Performance of African Countries on the Global (Human Capital) Indexes

It is of much concern that African countries have consistently been at the lower rung of the ladder in these rankings. Across all indexes, African countries constitute the bulk of the bottom 50 countries. Few African countries rank in the top hundred on almost all the indexes. When compared to countries with the same income levels from other world regions, the performance of African countries, overall, on the different indexes is rather poor. However, the major concern is not necessarily with the poor rankings of African countries when compared to countries with similar income levels from other regions. Rather, the concern is with the poor scorings of African countries across all the indexes and the various measurement indicators in the indexes. On the HCI edition of 2020, of the 46 African countries ranked, only 5 countries—Seychelles, Mauritius, Kenya, Algeria, and Tunisia—had a total score above 0.50 out of the available score of 1.0. On the Insead GTCI for 2023, no African country had a score above 50 points.

The state of a country's human capital does not only reflect the current state of a country's economic productive capacity; it also reveals and provides an insight into the potential future productivity of the country. For African countries, the rankings portend a somewhat gloomy picture of the future. The aspirations of Agenda 2063 of African Union, and any future similar visions, no matter how laudable may not be achieved if the right investments are not made in Africa's most valuable resources—the people—and if the right environment is not created for people to gainfully deploy their human capital. However, all hope is not lost. Fortunately, Africa has the youngest population and highest birth rates globally, although some seldom see the fortune in these, especially in the second factor, because of the inherent challenges that simultaneously correspond to these factors. Obviously, the best time for African countries to have invested in their human capital was in the past. The next best time is now, to invest in the young and growing youthful population of the continent.

Conclusion

The different global human capital indexes that have emerged over the years since the acceptance of the concept of human capital have great utility for countries despite the criticisms against them. The mini-case studies presented in the following three chapters, Chaps. 6, 7, and 8, were selected based on their overall performance on the three global indexes by global agencies: HDI, GHCI, and HCI. Their performance on Insead's GTCI was also considered in the selection. Each of the three chapters showcases the top country from different continental world regions across these indexes. Singapore was selected from Asia, Finland from Europe, and Canada from North America. These three countries have not only consistently ranked tops among their (geographical) continental peers; key aspects of human capital development and acquisition are also evidently shown in their economic progress and blueprint. Also, the three countries are showcased for their different and unique paths to human capital development and acquisition. Finally, Asia, Europe, and North America are the three top continents on the different indexes, showcasing countries with the most developed human capital globally. The geographical groupings across the three indexes do not align with one another. The groupings of countries on the index seem to align more along geopolitical zones rather than just geography. However, the selection of countries for the case studies in the subsequent chapters also depended majorly on geographical rather than geopolitical alignments.

References

Abraham, & Mallatt. (2022). Measuring human capital. *Journal of Economic Perspectives, 36*(3), 103–130.

Barro, R. J., & Lee, J. W. (1993). International comparisons of educational attainment. *Journal of Monetary Economics, 32*, 363–394.

Barro, R., & Lee, J. W. (2001). International data on educational attainment: Updates and implications. *Oxford Economic Papers, 53*(3), 541–563.

Barro, R., & Lee, J. W. (2013). A new data set of educational attainment in the world, 1950–2010. *Journal of Development Economics, 104*, 184–198.

Business Standard. (2018). *Rejecting human capital index, India says digital age demands better metric.* Business Standard. https://www.business-standard.com/article/economy-policy/rejecting-human-capital-index-india-says-digital-age-demands-better-metric-118101300545_1.html

Cohen, D., & Soto, M. (2007). Growth and human capital: Good data, good results. *Journal of Economic Growth, 12*, 51–76.

Corral, P., Dehnen, N., D'Souza, R., Gatti, R., & Kraay, A. (2021). The World Bank human capital index. In B. Fraumeni (Ed.), *Measuring human capital* (pp. 55–81). Elsevier.

Demirgüç-Kunt, A., & Torre, I. (2022). Measuring human capital in middle income countries. *Journal of Comparative Economics, 50*(4), 1036–1067. https://doi.org/10.1016/j.jce.2022.05.007

Economic Times. (2018, October 11). India rejects findings of World Bank report on human capital index. *The Economic Times.* https://economictimes.indiatimes.com/news/economy/policy/india-rejects-findings-of-world-bank-report-on-human-capital-index/articleshow/66170167.cms

Ederer, P. (2006). *Innovation at work: The European human capital index.* The Lisbon Council. Available at https://lisboncouncil.net/wp-content/uploads/2020/08/European-Human-Capital-Index.pdf

Ederer, P., Schuller, P., & Willms, S. (2007). *The European human capital index: The challenge of central and Eastern Europe.* A Lisbon Council Policy Brief. Available at https://lisboncouncil.net/wp-content/uploads/2020/08/European-Human-Capital-Index-CEE.pdf

Edwards, D. (2018). *What's wrong with the World Bank's human capital index?* Available at https://www.ei-ie.org/en/item/22632:whats-wrong-with-the-world-banks-human-capital-index-by-david-edwards

Esho, E., & Verhoef, G. (2020). A holistic model of human capital for value creation and superior firm performance: The strategic factor market model. *Cogent Business & Management, 7*(1). https://doi.org/10.1080/23311975.2020.1728998

Feenstra, R. C., Inklaar, R., & Timmer, M. P. (2015). The next generation of the Penn World Table. *American Economic Review, 105*(10), 3150–3182.

Fraumeni, B. M., & Liu, G. (2021). Summary of world economic forum, "the Global Human Capital Report 2017 – Preparing people for the future of work". In B. Fraumeni (Ed.), *Measuring human capital* (pp. 125–138). Elsevier.

GHCI. (2017). *The global human capital report: Preparing people for the world of work.* Insight Report, World Economic Forum (WEF).

GTCI. (2023). In B. Lanvin & F. Monteiro (Eds.), *The global talent competitiveness index: What a difference ten years make, what to expect for the next decade.* Insead Business School, France, & Human Capital Leadership Institute.

Hanushek, E. A., & Woessmann, L. (2012). Do better schools lead to more economic growth? Cognitive skills, economic outcomes, and causation. *Journal of Economic Growth, 17*, 267–321. https://doi.org/10.1007/s10887-012-9081-x

Hunter, B. M., & Shaffer, J. D. (2022). Human capital, risk and the World Bank's reintermediation in global development. *Third World Quarterly, 43*(1), 35–54. https://doi.org/10.1080/01436597.2021.1953980

Kim, J., & Mahoney, J. T. (2007). Appropriating economic rents from resources: An integrative property rights and resource-based approach. *International Journal of Learning and Intellectual Capital, 4*(1/2), 11–28. https://doi.org/10.1504/IJLIC.2007.013820

Lagakos, D., Moll, B., Porzio, T., Qian, N., & Schoellmann, T. (2014). *Experience matters: Human capital and development accounting*. National Bureau of Economic Research. Working paper 18602.

Lim, S. S., Updike, R. L., Kaldjian, A. S., Barber, R. M., Cawling, K., York, H., Friedman, J., et al. (2018). Measuring human capital: A systematic analysis of 195 countries and territories, 1990–2016. *Lancet, 392*, 1217–1234.

Liu, J. (2018). *Mind the learning gap: A methodological look into World Bank's new human capital index by Ji Liu*. Network for International Policies and Cooperation in Education and Training. https://www.norrag.org/mind-the-learning-gap-a-methodological-look-into-world-banks-new-human-capital-index-by-ji-liu/

Liu, G., & Fraumeni, B. M. (2020). *A brief introduction to human capital measures*. IZA Institute of Labor Economics, Discussion paper series, IZA DP No. 13494.

Liu, J., & Steiner-Khamsi, G. (2020). Human capital index and the hidden penalty for non-participation in ILSAs. *International Journal of Educational Development, 73*, 102149. https://doi.org/10.1016/j.ijedudev.2019.102149

Shukla, A. (2018). *What's the human capital index and why should policy makers take it seriously?* Available at https://gppreview.com/2018/11/28/whats-human-capital-index-policymakers-take-seriously/

Stein, F., & Sridhar, D. (2019). Back to the future? Health and World Bank's human capital index. *BMJ, 367*, l5706. https://doi.org/10.1136/bmj.l57064

Summers, R., & Heston, A. (1988). A new set of international comparisons of real product and price levels: Estimates for 130 countries, 1950–1985. *Review of Income and Wealth, 34*(1), 1–25.

UN. (2022). *Uncertain times, unsettled lives: Shaping our future in a transforming world*. Human development report 2021/2022.

Yusuf, S. (2020). *Building human capital, lessons from country experiences: How Singapore does it*. World Bank Report.

The Emergence of Asia: The Case of Singapore

6

Abstract

This chapter presents a case study on Singapore. Guided by the visionary leadership of Lee Kuan Yew, Singapore strategically used human capital to move from being a third world country in the 1960s to one of the most advanced countries in the world today. The case study highlights how the different phases of the economic and social growth of Singapore were wrapped around human capital development. This chapter also presents some of the human capital development programs that were, and continue to be, undertaken by Singapore to ensure that the country continues to acquire, accumulate, develop, and utilize human capital for sustained economic growth and development.

Keywords

Singapore · Economic transformation · Industrialization · Economic development · Human capital

Introduction

Like Africa, Asia is a diverse continent with different subregions and countries. However, unlike Africa, Asia is much more geographically dispersed with many countries amid oceans and islands. Many countries in Asia have emerged and transited from developing to developed countries. Countries such as Hong Kong, South Korea, Taiwan, and Singapore, sometimes referred to as the Asian tigers, in addition to Japan have emerged to become developed countries. Many other Asian countries such as China, India, Indonesia, Malaysia, the Philippines, Thailand, and Vietnam are also fast emerging and are likely to become categorized as developed countries in the near future. Indeed, China's gross domestic product (GDP) has overtaken that of many European countries and is now the second largest economy globally. Consequently, two of the top three economies globally, China and Japan, are geographically from Asia.

The major global human capital indexes, World Bank's Human Capital Index (HCI), United Nations' Human Development Index (HDI), and Insead Business School's Global Talent Competitiveness Index (GTCI), have differing categorization of Asian countries. However, it is still quite easy to see that Asia as a continent has made considerable progress in developing their human capital. Overall, Asian countries have achieved higher levels of human capital development than African countries. This progress is aptly evidenced in the economic and social progress that has been made in the region over the years. Some of the top Asian countries across the global human capital indexes over the years include Singapore, Japan, Hong Kong,

South Korea (Republic of Korea), Kazakhstan, Vietnam, China, and Malaysia in no particular order. Each of these countries can aptly serve as a case study of human capital and development.

However, Singapore, in particular, has made tremendous progress in strategically developing its human capital. The country has consistently ranked among the top countries globally on the HCI, HDI, GTCI, and World Economic Forum's (WEF) Global Human Capital Index (GHCI). Indeed, it ranked as the number one country in the maiden edition of HCI in 2018 and also in 2020 garnering a value of 0.88 in both years. By these rankings, it means that a child born in Singapore has the most chance to achieve full productivity potential and to be in full good health. The general success of many Asian countries in human capital development is not the result of following one peculiar or particular path. Different Asian countries have taken different approaches to human capital development (ADB, 2017; Baker & Holsinger, 1997). Singapore's approach to human capital development is therefore unique and not comparable to that of other Asian countries. The country has one of the highest levels of human capital development in the world, ranking among the top ten countries on all available indexes of human capital. According to World Bank's human capital reports, a child born in Singapore has a 100% chance of surviving till age 5 and can be expected to complete 13.9 years of schooling. This data reflects the efforts and strategies that have been put in place by Singapore to develop their human capital over the years.

Singapore: A Historic Economic Transformation

The South East Asian country of Singapore has made tremendous economic progress. Singapore has a total population that is currently below six million. Singapore initially gained independence from being a British colony in 1959 but became one country with Malaysia in 1963 (Cahyadi et al., 2004). Since gaining complete independence after separation from Malaysia in 1965,

Singapore has transformed into one of the countries with a captivating story of economic transformation. Data from the World Bank shows that the country's GDP grew from less than $12 billion in 1980 to over $96 billion in 2000 and to over $466 billion in 2022. At the country's independence, the GDP per capital was just merely above $500. To paint a clearer picture for comparison, in 1960, Singapore's GDP per capital was $428 which was less than the world average of $458 and less than one-sixth of that of the United States (Dreisbach, 2020). With a GDP per capita currently at over $80,000, Singapore is now one of the most prosperous countries in the world (World Bank, 2022a; Zhou, 2019). A Singaporean is, on the average, one of the most prosperous persons in the world (Fig. 6.1 presents the historic transformation in the county's GDP and GDP per capita). From a country riddled with high unemployment, and ethnic tensions, Singapore's economy has transformed into one of the most developed economies in the world.

Singapore consistently ranks among the top countries in many global indexes of competitiveness and innovation. In the most recently released World Competitiveness ranking by the Institute of Management Development (IMD) Switzerland, Singapore ranked in fourth place.[1] Although global rankings have limitations that one must be weary of, the top rankings of Singapore across many performance indexes are reflections of the economic development and outcomes of right policy decisions and implementation. The country can boast of being one of the countries with the highest standard of living globally. Lacking in natural mineral resource endowments, deliberate and strategic human capital development has been one of the major bedrocks of the country's economic development and transformation. For instance, as part of education reform in 1992, every student in Singapore was required to have a minimum of 10 years mainstream education (OECD, 2010; Yang, 2022). In fact, Singapore explicitly considers its people as its most strate-

---

[1] https://worldcompetitiveness.imd.org/countryprofile/ overview/SG

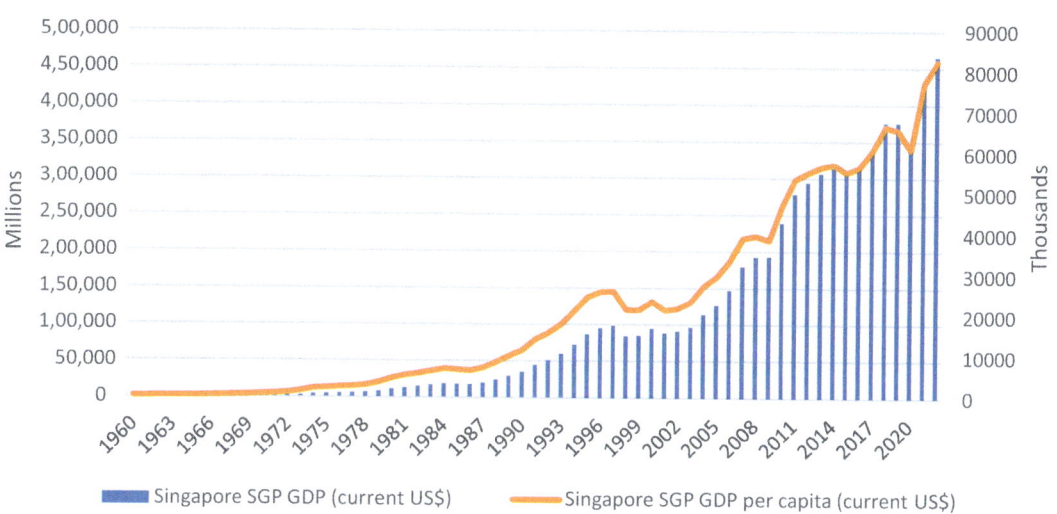

Fig. 6.1 Singapore's GDP and GDP per capita in current US$. (Source of data for graph: World Development Indicators (2024a, b))

gic resource (Osman-Gani, 2004) and goes as far as providing the country's own definition and dimensions of human capital (Leggett & Cook, 2014). In a recent state visit to Kazakhstan in May 2023, Halimah Yacob, the President of Singapore, re-emphasized the country's commitment to human capital development as the key to its sustainable growth (Shayakhmetova, 2023).

The story of Singapore's economic transformation cannot be told without making reference to Lee Kuan Yew, the country's founding Prime Minister who was at the helm of governance from 1959 to 1990. Admirers, and critics alike, of his governance, pay glorious tributes to his transformational role of Singapore from a third world country to what it is today. To say the least, his governance was effective regardless of the voice of critics largely from upholders of liberal democracy. Indeed, Lee Kuan Yew's successful leadership of Singapore brings to fore the debate of whether the tenets of liberal democracy, as practiced by the west, are suited for all countries. Lee Kuan Yew's leadership was primarily based on meritocracy, the rule of law, and transparency with a high penalty for corruption (Reza et al., 2015; Quah, 2022; Yusuf, 2020). He himself, along with other government officials, eschewed corruption. With zero tolerance for corruption anchored on a strong political will, he spear-

headed Singapore's corruption framework which rests on four pillars: effective laws, an independent judiciary, effective enforcement, and a responsive public service (CPIB, n.d.; Quah, 2022). Despite the weaknesses of his governance style and policies, its results are, without doubt, globally evident.

Singapore's economic transformation is an example of the economic possibilities that can be realized when countries look beyond their natural resources and take deliberate strategies to advance the economy with human capital development at the core. From the low-skilled labor-based manufacturing of the 1960s and 1970s, Singapore moved to one based on high-level skills in the later decades of 1990 and beyond (Cahyadi et al., 2004; Yusuf, 2020). Industrialization in Singapore has advanced to value-added manufacturing in electronics, precision engineering, and chemicals and biomedical sciences (World Bank, 2019; Yeo, 2016). Through its commitment to human capital development and its effective utilization and deployment, manufacturing and services remain the country's striving industries having crafted its global competitiveness around human capital development.

A Phased Transformation Around Human Capital

The categorization that follows below breaks the historical development of Singapore into three phases to reveal a transformation wrapped around human development. Other writers have also separated Singapore's economic history and historical development since independence into different phases (e.g., Bozok, 2023; Cahyadi et al., 2004; Tan & Low, 2016; The Straits Times, 2022; Yeo, 2016; Yusuf, 2020). However, some attribute an emphasis on human capital development only to some phases (e.g., Yeo, 2016). However, national policies in all the phases of Singapore's economic development, regardless of time period of categorization and how it is broken down, were actually wrapped around human capital. In recognition of this, Tan and Low (2016), for example, used the phases of education in Singapore and categorized its economic history into four phases: survival-driven, efficiency-driven, ability-driven, and student-centric phases. This categorization by Tan and Low (2016), professors in Singapore (Rajandiran, 2020), is indeed a reflection of the role human capital has played in the development of Singapore. Although excellent physical infrastructure and necessary formal institutions were put in place, to attract foreign direct investments (FDI) in the early decades after independence, the state of Singapore's human capital in the 1960s was the key consideration for its national policy of low-skills industrialization. In other words, the type of FDI attracted was hinged on the type of human capital available in the country at the time. Indeed, in later decades, and even going further into the country's plans for the future, human capital remains the most important factor considered in crafting national policies.

Phase 1: Low-Skills-Based Industrialization At independence, a major challenge for the new government under the auspices of People's Action Party (PAP), led by Lee Kuan Yew, was high unemployment. In addition to this, most Singaporeans had little or no skills. Consequently, the general stock of human capital of Singapore was very low. Workers' conflict and labor strikes were also endemic (Bozok, 2023; Yusuf, 2020; Kanchoochat, 2019). The government's strategy, with advice from United Nations economic advisors such as Dr. Albert Winsemius, was to focus on attracting FDI, using the auspices of the Economic Development Board (EDB) which was established in 1961 (Cahyadi et al., 2004; Hosono, 2022; Kanchoochat, 2019). Ab initio, EDB aimed to promote industrialization in Singapore by establishing new industries and accelerating the growth of new ones (Hosono, 2022). However, given the state of its human capital, the government chose to attract low-skilled manufacturing companies. This was an excellent decision as ample current empirical research show that FDI usually flow to countries with the requisite human capital (Anetor et al., 2023; Asongu et al., 2018; Filippaios et al., 2019).

Oftentimes, the research information of the importance of human capital to FDI is interpreted to mean that to attract FDI, countries need to have a highly skilled workforce. However, this is an incorrect interpretation, and Singapore's application of this fundamental theoretical proposition in its early historical context is sample and ample proof of this.[2] Singapore focused on attracting the type of FDI that could make optimal use of the low level of skills available in the country as at the time. This is in no way discounting the other factors that were put in place to attract foreign investments. Industrial estates such as the Jurong Industrial Estate, for example, and state-owned enterprises were established in this period (Kanchoochat, 2019; Yeo, 2016). There are multiple factors that attract FDI into countries. However, recipient countries benefit most from FDI when the required human capital are available within the country. The attraction of FDI and the derivation of benefits from FDI are

---

[2]The government of Singapore may not have made this decision based on expert advice from research; scientific research, especially empirical-based research, sometimes only brings to fore principles and theories that are already inherently available. Scientific research offers the proof and validation required for some extant theories and principles.

two very distinct issues. There needs to be a proper matching of the human capital available in a country with the FDI that is attracted into the country for there to be a substantial impact in the economy and in the lives of people. Singapore did this majorly in the 1960s and into the 1970s.

However, the government did not rest on just attracting the FDI that matched the available stock of human capital. To meet the gap in skills, various human capital investments were made in different areas. The Singapore Vocational Institute was established in 1964, and a technical education department was established in 1968 to oversee the development of technical education, industrial training, and professional teachers' training (Low, 2002; Yang, 2022). Budgetary allocation to education increased seven times between 1960 and 1963, and the number of people with access to formal education increased by 50% (Bozok, 2023). Prior to independence, formal education was a privilege enjoyed only by a few—the wealthy. While low-skills-based industrialization was going on, large investments were concurrently made in human capital development to fill existing gaps in human capital.

Phase Two: The Skills-Building Phase This phase can be said to have started in the late 1970s and extended into the 1990s. In this phase, Singapore continued to attract FDI into other sectors other than low-skilled manufacturing. As the level of human capital increased in the country, the FDI attracted changed to ensure there was a continual match between FDI and human capital. Companies in the then emerging electronics industry in Europe, the United States, and North Asia were encouraged to extend their operations to Singapore and to set up their research and development arms in the country (Yeo, 2016). This period spanned between the 1970s and the 1980s when the country attracted an average of 2.3 billion dollars per year in FDI (Mirza, 2011; Yang, 2022). The influx of multinational companies into Singapore during this phase, in addition to those that were already in the country in the prior phase, brought along with it a correspond-

ing influx of foreign skilled workers (Low, 2002; Yusuf, 2020).[3]

To benefit from the immigration of foreign workers and to also enable knowledge transfer at this phase, Singapore began strategic investments in its people. Without this, the country would have continued to depend on low-skills industrialization while the state of technological advancements, which was beginning to gain grounds, continued globally. In other words, having multinational companies operating within Singapore was not enough. Singaporeans also had to be gainfully employed in these multinational companies and the many domestic companies emerging in the various value chain activities around the multinational companies. To achieve this, there was a need to make investment in a skilled local workforce. So, while there was an upscale in the types of FDI inflows into Singapore, there was a corresponding investment in increasing the human capital of the populace from low skills to semi- and high skills.

Having succeeded in attracting FDI that could take advantage of the skills level in the country, the focus shifted to deliberately building the skills of the people to higher levels. This phase saw many policies and programs aimed primarily at increasing the skills level of human capital available in the country. To achieve this, various education policies and national agencies were either established or merged to guide human capital acquisition by Singaporeans. In addition to the secondary vocational schools that had been established in the prior phase, postsecondary technical and vocational educational institutions were also established to equip people with the necessary technical skills (Yang, 2022). Subsequently, vocational institutes gradually replaced the secondary vocational schools (Garavan et al., 2016; Yang, 2022). Consequently, we see a deliberate human capital development strategy consisting of blending the twin tools of

---

[3]Attraction of foreign workforce has continued till date and currently accounts for about a third of the workforce in Singapore (Yang, 2022; Queux & Kuah, 2020), although now mostly made up of low-skilled workers.

formal education and foreign-trained personnel from the multinational companies that had been attracted to Singapore. Among other benefits, this strategy facilitated and eased knowledge transfer between foreign skilled workers and Singaporeans.

Singapore's financial sector blossomed as it provided much support to the much successful industrialization. By the 1980s, the financial sector became one of the largest sectors in the economy and contributed as much as 30% to the country's GDP growth in 1983 (SG101, n.d.). The country's export-led industrialization through multinational companies had been hugely successful. Singapore had become a forefront country amid developing countries in Asia and around the world, leaving behind other developing countries in Africa and Latin America that had been in the same position with them in the 1960s.

Phase 3: Highly Skilled Industrialization By the 1980s and 1990s, it was becoming quite obvious to the world that Singapore was making good economic and social progress. Singapore was no longer the poor group of islands with little or no natural resource. With few exceptional years, GDP growth rate for Singapore hovered around 8% per year (World Bank, 2022b), and per capital income had now increased to about $13,000 (Menon, 2015). Life expectancy at birth which was about 65 years in 1960 had increased to 77 years in 1998 (Cahyadi et al., 2004).

A key component of this next phase was the diversification of the economy. Industrial parks were built to support the growth of biomedical and biopharmaceutical industries, and the focus of industrialization shifted to a "knowledge-intensive industrial structure" (Hosono, 2022; Siddiqui, 2010). Consequently, the focus of the government was to diversify the economy away from the financial sectors and already existing manufacturing industries and to gear the economy toward newly emerging global industries in electronics, computer sciences, biomedical and biopharmaceutical industries, and similar knowledge-based industries. The National

Science and Technology Board (NSTB) was established in 1991, and the National Computer Board was created to provide incentives to attract FDI into the electronic sector (Siddiqui, 2010). Universities entered into strategic global partnerships such as research collaborations with foreign universities for advancement in research and technology (Cahyadi et al., 2004; OECD, 2010). Investment in education during this phase went beyond technical and vocational education that had been the focus of the government in the previous phase. Science and technology became the main focus, and corresponding agencies, policies, and programs were put in place to develop the necessary skills even as the government continued the deliberate attraction of multinational companies in electronics, science, and high-tech industries. At the end of this phase, Singapore had successfully transformed itself into a developed country.

Continuous Human Capital Investments into the Future

The above three phases can hardly be categorized into discrete time periods. Each phase overlaps with the next. The categorization into three phases has been for ease and to clearly show the role human capital has played in the national development of Singapore. To underscore its commitment to human capital development from the onset, 40% of the $870 million (26% to education and 14% to health services and public health) dedicated to the first national development plan was budgeted for human capital development (Dreisbach, 2020; Lee, 2018; Lim, 2017). Since the end of the third phase, Singapore has continued to invest in developing its human capital. In 2001, in preparation for the future, Singapore's ministry of manpower released a statement specifying four different focus areas of its human capital development—knowledge capital, imagination capital, emotional capital, and social capital (Leggett & Cook, 2014). Whether or not one agrees with this typology of human capital is irrelevant. This attempt at a typology of human capital typifies a country that is deliberate enough to explicitly specify its understanding of human capital. Beyond this, it also signifies a country that understands that its number one resource is its people and the human capital: knowledge, skills, and abilities, embedded in them as individuals. Singapore seems to have understood and continues to understand that a country can be built on the backbone of people's human capital.

Human Capital Development Programs

On a cursory look, Singapore's strategic development may at first seem to have been solely anchored principally on attracting FDI. However, a closer look reveals that human capital development has always been at the core of the trajectory of the country's economic development. All through the different phases of Singapore's developmental history, a key question was and continues to be whether Singaporeans have the necessary knowledge and skills for the FDI attracted. In addressing the skills gap in Singapore, the government has never relied on a single policy. It has always employed diverse policies in its bid to develop its human capital. Formal tertiary education, for example, does not revolve only around college or university education. Different functional technical and vocational colleges and institutions were set up at various stages to meet the needs of the country. This was mingled with appropriate immigration policies to attract the necessary skills not available in the country to meet the needs of the many multinational companies that were coming into the country.

National policies for human capital development were and still are deliberate and coordinated. The immediate and long-term human capital needs of the country are assessed with the aim of taking actions on how to fill the identified gaps in conjunction with universities, polytechnics, and other formal tertiary educational institutes (Osman-Gani, 2004). Singapore continues to project the skills that may be required for work in the future and provides the necessary invest-

ments and policies to ensure that the country is able to meet its requirements in the future. For example, the Workforce Development Agency (WDA) was established in 2003 to enhance the employability and competitiveness of workers and job seekers to ensure that the workforce continues to meet the changing needs of Singapore (Leggett & Cook, 2014; Osman-Gani, 2004). Singapore's commitment to human capital development is also revealed in the several human capital development agencies and programs that have been put in place over the years. A tripartite structure ensures that the government; workers, through the labor unions; and the private sector have a place and role in the continuous development of the country's people. Several private sector-led programs also exist to enable people acquire human capital. These private-led programs are conducted in conjunction with foreign universities and provide opportunities to those unable to go through the formal education system (Osman-Gani, 2004). Below are some national programs and initiatives that reflect this tripartite relationship and the government's commitment to human capital development.

The Teacher Education Model: TE21 To underscore the importance placed on human capital, in 2009, Singapore introduced a teacher education model. Named Teacher Education Model for the 21st Century (TE21), it was set up to equip teachers with twenty-first-century competencies that are able to meet the knowledge-driven and fast-evolving technologies of the times. TE21 is geared toward a quality education system with the capacity to produce students that meet the requirements of the twenty-first century both locally within Singapore and globally. TE21 recognizes that "the quality of the teaching force determines the quality of education" (NIE, 2010: p. 1). Under TE21, teaching is seen as a profession that requires a degree as a prerequisite for joining the profession (NIE, 2010). The model aims to produce professional teachers with the values, knowledge, and skills to meet the challenges of teaching in the twenty-first-century classroom.

The Skills Development Fund (SDF) The SDF was created in 1979 to encourage employers in Singapore to train their employees. The fund is funded by a levy and provides "financial incentives for persons preparing to join the workforce, persons in the workforce and persons rejoining the workforce" (UNESCO, 2022: p. 126). SDF can be regarded as an employer contributory scheme for employee development initiative. Employers contribute a percentage of the wages of their employees to the fund and can apply for skills-training grants to enable them train their employees (Leggett & Cook, 2014; Osman-Gani, 2004). This has served as an effective means of motivating employers, who otherwise may not have been inclined to develop their employees, to provide training and development to their employees. Employers can send their employees to the training programs provided by the SDF, thereby taking advantage of their contributions to the fund, and receive skills-training grants or loans from SDF (Leggett & Cook, 2014; Osman-Gani, 2004; UNESCO, 2022). Eligible employers who provide in-house training to their employees can also apply for subsidies to reduce the cost of such trainings (UNESCO, 2022).

The SkillsFuture Initiative This was launched in 2014 by the Workforce Development Agency (WDA) and kicked off in 2015 to build a culture of developing skills and lifelong learning (Ang, 2021). The SkillsFuture Initiative aims to encourage and enable Singaporeans to develop skills relevant to the future and take Singapore to the next level of economic development (Tan, 2017). The hope is that through this forward-looking initiative, Singaporeans can go beyond the acquisition of paper certificates to learning and expertise that advances innovation through a focus on mastery, meritocracy, and the individual's interests, aspirations, and passions (Tan, 2017). Through this initiative, all Singaporeans who are 25 years old and above can access $500 to pay for a broad range of approved courses (Tan, 2017). This initiative is a subtle recognition of the role human capital has played in bringing Singapore to its current level of development. Singapore, through

this initiative, shows it is not resting on its oars. The aspiration for continuous development through human capital development is clear.

Apprenticeship Training Singapore has a formal apprenticeship training that is similar to Germany's dual training system. School leavers have the option of applying for the several apprenticeship programs provided by the Institute of Technical Education (ITE), on-the-job with employers and off-the-job at the ITE (Osman-Gani, 2004).

The Human Capital Partnership (HCP) Program Singapore's HCP program was launched in February 2017 to encourage employers to take active steps in investing in human capital. Through HCP, employers are encouraged to commit to making investments in their employees at all levels, build complementarity between local and foreign manpower, and enhance knowledge transfer from foreign workers to local workers in their employ. By becoming enlisted in the HCP program, organizations are obligated to have human resource (HR) policies and practices that ensure that employees' training and development becomes a priority. Companies on the HCP program receive a special visual HCP mark that designates them as HCP partners (Tripartite Alliance, n.d.). This visual symbol helps to identify these organizations as exemplary employers, helping them to attract talented employees.

Singapore's Central Provident Fund The Central Provident Fund (CPF) is not exactly a human capital development fund. However, it can be regarded as one because it has helped to fund aspects of Singaporean's education and health, two critical components of human capital, and for health-care strategy. Established in 1955, the CPF is a social security scheme providing funds for housing purchases and medical and other benefits (Cahyadi et al., 2004). The CPF is compulsory for all working individuals in Singapore (Cahyadi et al., 2004).

Health-Care Strategy

Health is an inherent fundamental component of human capital that needs to be planned for economic productivity. Without good health, human capital is incomplete, and deployment in production is hampered. Provision for health-care services and public health was a priority for the country from the onset of the first national development plan in 1961. Being a densely populated country situated on only 716 square kilometers (Dreisbach, 2020), provision for health invariably included urban planning, provision of housing, and waste management as these are factors that affect quality of health. The health-care strategy involved integrating housing and urban development plans. The Housing Development Board (HDB) set up in 1960 to provide housing is therefore an integral part of the health-care strategy of Singapore. As at 2013, 85% of Singaporeans lived in houses provided by the HDB (Dreisbach, 2020). Clinics and outpatients dispensaries, particularly for maternal and child care, set up in the 1960s as one-stop medical centers to cater for various medical needs, have evolved into modern well-equipped medical centers (Haseltine, 2013). Although initially set up as free health-care centers, they quickly changed to fee paying to curtail wastage and sustain quality. The committee for postgraduate medical education set up in 1970, in collaboration with medical institutions around the world, has helped to boost the number of specialist medical personnel (Haseltine, 2013).

Human Capital Development Lessons from Singapore

The dividends of a human capital-based approach to economic development in the country have been enormously visible for all to see. There was a deliberate attraction of FDI into its manufacturing industries in the early two decades after independence and later into the services industry. However, this was aptly anchored on human capital. Policies such as the establishment of the National Trade Union Congress, a single trade union established to oversee all employment and wage problems in 1972 and the National Computer Board in 1981 (Cahyadi et al., 2004), were implicitly centered around people at the core. Consequently, national policies were ultimately centered around an understanding of the people and their skills. Through various programs and national structures put in place for formal education and for organizations to train employees, Singapore appears to have a firm grasp of the meaning and nature of human capital.[4] While attracting FDI played a role in its human capital development, domestic resources were also mobilized (Yusuf, 2020). So, the focus was not merely on attracting FDI.

What was also paramount was access to and effective utilization of the necessary resources. Singapore has consistently prepared for the future by investing in the knowledge and skills that could be required in the future in its people. However, it has also always taken cognizance of the country's shortfalls and created the right environment to bring in foreign investments and skilled people required to advance the economy. In crafting a human capital-based national strategy, there was a holistic integration of policies and programs directed toward a common national goal that was considerably well understood. Consequently, there was little room for silos within the various policies and government agencies and institutions.

Several other lessons can be gleaned from the "Singapore" story, especially when a more in-depth analysis of the various national policies and structures put in place to support the overall national strategy is conducted. The inclusion of private sector and employees in programs and initiatives and various tripartite arrangements in various programs ensured that members from various parts of the society were incorporated into programs and initiatives. In addition to these, Yusuf (2020) identifies the collection and analysis of data for policy decisions, able and incorruptible leaders with high standards, and the establishment of a merit-based and non-

---

[4] Please refer to Part 1 of the book for the definitions, meanings, and nature of human capital.

politicized bureaucracy as three main factors that has undergirded Singapore's successful implementation of a human capital-based strategy. However, just like Singapore did not rely on any "copy and paste" strategy from other countries, it would be fool hardy to wholly adopt Singapore's strategy without cognizance of peculiar national contexts. Nonetheless, for African countries, the case of Singapore illustrates how developmental policies can be anchored on people and their human capital. For instance, the drive to attract FDI inflows need to be tailored to the peculiarities of each African country while taking cognizance of the current levels and nature of available stock of human capital. The story of Singapore also underscores the importance of combating corruption and related vices and the utility of meritocracy and accountability of public office holders for economic growth and development.

Conclusion

Singapore has demonstrated that strategic investments in human capital can help achieve national social and economic development. With little natural mineral resources, Singapore seemed to have no comparative advantage in relation to other nations. However, with a developmental strategy crafted around human capital development, foreign and domestic investments were used to create a comparative advantage and to move the economy from one that was struggling in the 1960s to one of the most developed economies today. Singapore's transition from an under developed economy to an advanced one was shaped by the dominant roles of government and public sector. Private sector's role was trimmed to that of business, though well incorporated into the national programs and initiatives. Any private sector-led human capital developmental efforts were primarily done by ample practical encouragement of the government. Some have argued that Singapore's relatively small population and landmass size aided the implementation of national policies. While this may have an element of truth, the significant factor has been the strategic focus on human capital development. Another key factor has been the coordinated implementation of the whole human capital-focused strategy. With the lessons properly gleaned from the case of Singapore, other nations could potentially have similar success as Singapore. Adequate focus needs to be placed on, and national strategies need to be correctly wrapped around, human capital development.

Singapore, a country without the abundant natural resources available to many African countries, has managed to become one of the most competitive and developed countries in the world. The story of Singapore's industrialization and economic advancement is incomplete without an acknowledgment of the strategic role of human capital. Although not always explicitly expressed, or even intentionally done, Singapore's developmental path conforms with an approach that aligns with a strategic human capital approach to long-term and sustainable development advocated for in this book.

References

ADB. (2017). *Human capital development in South Asia: Achievements, prospects, and policy challenges.* Asian Development Bank Report.

Anetor, F. O., Aziegbe-Esho, E., & Verhoef, G. (2023). Do the quality of governance, levels of human capital and financial development matter for foreign direct investment flows to Africa? *Interdisciplinary Journal of Economics and Business Law, 12*(4), 34–63.

Ang, J. (2021). 540,000 Singaporeans benefitted from the SkillsFuture Initiatives in 2020. *The Straits Times.* https://www.straitstimes.com/singapore/540000-singaporeans-benefited-from-skillsfuture-initiatives-in-2020-skillsfuture-singapore

Asongu, S., Akpan, U. S., & Isihak, S. R. (2018). Determinants of foreign direct investment in fast-growing economies: Evidence from the BRICS and MINT countries. *Financial Innovation, 4*(26), 1–17. https://doi.org/10.1186/s40854-018-0114-0

Baker, D. P., & Holsinger, D. B. (1997). Does a unique regional model exist? In D. Bradshaw (Ed.), *Education in comparative perspective: New lessons from around the world* (International Studies in Sociology and Social Anthropology Series, 63, Brill) (pp. 159–173).

Bozok, F. (2023). *The impact of human capital development and education on Singapore's economic success.* Economic History Blog, Social Sciences University of Ankara. Available at https://medium.com/economic-history-blog/the-impact-of-human-capital-development-and-education-on-singapores-economic-success-517463406b39

Cahyadi, G., Kurseten, B., Weiss, M., & Yang, G. (2004). *Singapore's economic transformation*. A Global Urban Development Report.

CPIB. (n.d.). *Singapore's corruption control framework*. Available at https://www.cpib.gov.sg/about-corruption/prevention-and-corruption/singapores-corruption-control-framework/

Dreisbach, T. (2020). *Integrating human capital into national development planning in Singapore*. Delivery Note, Global Delivery Initiative of World Bank's Human Capital Project.

Filippaios, F., Annan-Diab, F., Hermidas, A., & Theodoraki, C. (2019). Political governance, civil liberties, and human capital: Evaluating their effect on foreign direct investment in emerging and developing countries. *Journal of International Business Studies, 50*, 1103–1112.

Garavan, T. N., McCarthy, A. M., & Morley, M. J. (2016). *Global human resource development: Regional and country perspectives*. Routledge.

Haseltine, W. (2013). *Affordable excellence: The Singapore healthcare story*. Ridge Books/Brooking Institution Press. https://www.brookings.edu/wp-content/uploads/2016/07/AffordableExcellencePDF.pdf

Hosono, A. (2022). *SDGs, transformation, and quality growth: Insights from international cooperation*. Springer Nature.

Kanchoochat, V. (2019). Tigers at critical junctures: How South Korea, Taiwan and Singapore survived growth-led conflicts. In Y. Takagi, V. Kanchoochat, & T. Sonobe (Eds.), *Developmental state building. Emerging-economy state and international policy studies*. Springer. https://doi.org/10.1007/978-981-13-2904-3_3

Lee, J. (2018). *S'pore's first-ever budget in 1965 splurged on education & hospital services*. Mothership. https://mothership.sg/2018/02/singapore-first-budget-1965/

Leggett, C., & Cook, J. (2014). Human capital development in Singapore. In M. Gunderson & F. Fazio (Eds.), *Tackling youth unemployment* (Adapt labour studies book series). Cambridge Scholars.

Lim, P. (2017). *State development plan, 1961–1964*. Singapore Infopedia. https://eresources.nlb.gov.sg/infopedia/articles/SIP_2017-10-11_092937.html

Low, L. (2002). The political economy of migrant worker policy in Singapore. *Asia Pacific Business Review, 8*(4), 95–118.

Menon, R. (2015). *An economic history of Singapore: 1965–2065*. An address by the Managing Director of the Monetary Authority of Singapore, at the Singapore Economic Review Conference 2015, Singapore, 5 August 2015.

Mirza, H. (2011). *Multinationals and the growth of the Singapore Economy*. Routledge.

NIE. (2010). *A teacher education model for the 21st century (TE21): A report by the National Institute of Education, Singapore*. Available at https://www.nie.edu.sg/docs/default-source/te21_docs/te21_executive-summary_14052010%2D%2D-updated.pdf

OECD. (2010). *Singapore: Rapid improvement followed by strong performance*. Available at https://www.oecd.org/countries/singapore/46581101.pdf

Osman-Gani, A. M. (2004). Human capital development in Singapore: An analysis of national policy perspectives. *Advances in Developing Human Resources, 6*(3), 276–287.

Quah, J. S. T. (2022). Lee Kuan Yew's role in minimizing corruption in Singapore. *Public Administration and Policy: An Asia-Pacific Journal, 25*(2), 163–175.

Queux, S., & Kuah, A. T. H. (2020). Junzi leadership in Singapore: Governance and human capital development. *Journal of Management Development, 40*(5), 389–403.

Rajandiran, D. (2020). Singapore's teacher education model for the 21st century (TE21). In F. M. Reimers (Ed.), *Implementing deeper learning and 21st century education reforms* (pp. 59–77). Springer. https://doi.org/10.1007/978-3-030-57039-2_3

Reza, S. M. S., Raihan, S. M. S., & Rabi, M. R. I. (2015). Human resource management and leadership: Lessons from Lee Kuan Yew. *International Journal of Business and Management Study, 2*(2), 268–271.

SG101. (n.d.). *1986 to 1996: Rebounding into a decade of growth*. Available at https://www.sg101.gov.sg/economy/growing-our-economy/1986/

Shayakhmetova, Z. (2023). *Singapore and Kazakhstan are forward-looking nations committed to human capital development, sustainable growth, says Singapore's President*. Online news report available at https://astanatimes.com/2023/05/singapore-and-kazakhstan-are-forward-looking-nations-committed-to-human-capital-development-sustainable-growth-says-singapores-president/

Siddiqui, K. (2010). The political economy of development in Singapore. *Research in Applied Economics, 2*(2), E4.

Tan, C. (2017). Lifelong learning through the SkillsFuture movement in Singapore: Challenges and prospects. *International Journal of Lifelong Education, 36*(3), 278–291. https://doi.org/10.1080/02601370.2016.1241833

Tan, O., & Low, E. (2016). Singapore's systematic approach to teaching and learning twenty-first century competencies. In F. M. Reimers & C. K. Chung (Eds.), *Teaching and learning for the twenty-first century: Educational goals, policies, and curricula from six nations*. Harvard Education Press.

The Straits Times. (2022). *The education system over the years*. Available at https://www.straitstimes.com/singapore/education/the-education-system-over-the-years

Tripartite Alliance. (n.d.). *Being exemplary: Learning about the Human Capital Partnership Programme (HCP)*. https://www.tal.sg/tafep/getting-started/exemplary/hcp-programme#

UNESCO. (2022). *Global review of training funds; Country Briefs: Singapore*. Available at https://unevoc.

unesco.org/countryprofiles/docs/UNESCO_Funding-of-Training_Singapore.pdf

World Bank. (2019). *The world bank in Singapore: Overview.* https://www.worldbank.org/en/country/singapore/overview#:~:text=In%20the%20most%20recent%20World,complete%20education%20and%20full%20health

World Bank. (2022a). *GDP per capita (Current US$) – Singapore.* Available at https://data.worldbank.org/indicator/NY.GDP.PCAP.CD?locations=SG

World Bank. (2022b). *GDP growth (annual %) – Singapore.* Available at https://data.worldbank.org/indicator/NY.GDP.MKTP.KD.ZG?locations=SG

World Development Indicators. (2024a). *GDP growth (annual %) – Singapore.* Available at https://data.worldbank.org/indicator/NY.GDP.MKTP.KD.ZG?locations=SG

World Development Indicators. (2024b). *GDP per capita (Current US$) – Singapore.* Available at https://data.worldbank.org/indicator/NY.GDP.PCAP.CD?locations=SG

Yang, L. (2022). The human resource development policy of Singapore. *Academic Journal of Humanities & Social Sciences, 4*, 86–88.

Yeo, P. (2016). Going beyond comparative advantage: How Singapore did it. In R. Cherif, F. Hasanov, & M. Zhu (Eds.), *Breaking the oil spell: The Gulf Falcons' path to diversification.* International Monetary Fund (IMF). https://www.elibrary.imf.org/display/book/9781513537863/ch004.xml

Yusuf, S. (2020). *Building human capital, lessons from country experiences: How Singapore does it.* World Bank Report.

Zhou, P. (2019). *The history of Singapore's economic development.* Available at https://www.thoughtco.com/singapores-economic-development-1434565

Integrated Europe: The Case of Finland

7

Abstract

This chapter presents the case study of Finland, a Nordic country in Northern Europe that has performed very well on many human capital indexes and other global indexes. This chapter focuses on the education system of Finland which has been proclaimed by many as one of the best in the world due to the success of Finnish students on Organisation for Economic Co-operation and Development (OECD)'s Programme for International Student Assessment (PISA). Undoubtedly, the education system in Finland has many features that are uncommon in many other parts of the world. However, the current renowned education system is not the product of these features but results from historical reforms that dates back to the eighteenth century and rests on some fundamental principles derived from the Nordic welfare system. This chapter highlights the reforms and the historical contexts of Finland's education system, as well as the health-care system, as they form the backbone of the country's human capital development. The human capital development lessons from Finland, Nordic welfare states, and Europe in general conclude this chapter.

Keywords

Education system · School reforms · Compulsory schooling · Quality teaching · Human capital

Introduction

Unlike Africa, the continent of Europe is much more geographically dispersed. Europe is also more economically and politically integrated through the European Union (EU) than Africa, although not all European countries belong to the EU bloc. When it comes to human capital development, Europe can be regarded as one of the most advanced continents. Overall, Europe is second only to North America in human capital development on the different global human capital indexes. Many European countries usually rank among the top countries on the indexes for human capital. On the first edition of World Bank's Human Capital Index (HCI) in 2018, for example, 14 of the top 20 countries on the index were from Europe. Similarly, on the 2020 edition of HCI, 12 European countries were ranked among the top 20 countries on the index. This trend cuts across other global human capital indexes such as the United Nations' Human Development Index (HDI), World Economic Forum's (WEF) Global Human Capital Index (GHCI), and Insead's Global Talent Competitiveness Index (GTCI). European countries constitute over half of the top 50 countries on all the global human capital indexes. European countries are also conspicuously missing from among the least performing countries across all global indexes of human capital.

Compulsory Schooling and School Reforms in Europe

Some historical factors account for this general good performance of European countries across these indexes. One of them is the compulsory school reforms that took place across Europe in the mid-nineteenth and twentieth centuries (Hippe, 2020; Gathmann et al., 2012; World Bank, 2018). Literacy laws by formal and informal institutions such as church laws made it compulsory for citizens in many European countries to learn to read (Hippe, 2020). Some Scandinavian countries with mainly Protestant societies passed these compulsory literacy laws before the nineteenth century. The church law passed in the Kingdom of Sweden in 1686, for example, made it compulsory for people to learn to read (Hippe, 2020). Although some of these compulsory literacy laws were religious,[1] they, nevertheless, enabled reading and literacy that became beneficial beyond religion; many of these compulsory literary laws had subsequent unintended economic benefits. As Becker and Woessmann (2009) show in their analysis, these and other church laws helped in the formation of human capital in many European countries.

Beginning from the eighteenth century in Prussia and in the nineteenth century in many European countries, formal school attendance became compulsory when the first compulsory schooling laws were introduced (Gathmann et al., 2012; Hippe, 2020). Countries such as Denmark, England and Wales, the Netherlands, and Belgium, for example, passed their initial compulsory schooling laws in 1814, 1880, 1900, and 1914, respectively (Gathmann et al., 2012; Hippe, 2020). Over several subsequent years, many European countries instituted and implemented other compulsory schooling reforms, apart from the initial ones. These reforms inevitably helped many European countries to lay good foundations for human capital development.

Scandinavian and Nordic countries, in particular, have benefitted from these laws and reforms and generally rank well across global human capital indexes. Remarkably, the Nordic countries, Norway, Sweden, Denmark, Finland, and Iceland, all rank among the top countries across all global human capital indexes. On Insead's Global Talent Competitiveness Index of 2022 ranking of 133 countries, for example, Denmark, Sweden, Norway, Finland, and Iceland, ranked at the 3rd, 5th, 7th, 8th, and 12th positions, respectively. All but one Nordic country made it to the top ten ranked countries on the index. Indeed, all 5 countries rank among the top 30 countries in many global indexes such as the Global Competitiveness Index, Ease of doing business, Best Countries for Business, Index of Economic Freedom, Global Entrepreneurship Index, Social Progress Index, and the Corruption Perception Index (Deloitte Insights, 2020). It is, therefore, accurate to conclude that there is something that all Nordic countries have done and are possibly still doing, both in terms of talent and human capital development and other developmental areas, that is working effectively, at least for their countries' national contexts.

The compulsory school reforms in Europe also helped to establish formal school systems, especially basic primary and secondary school systems. There are ample evidences of formal education from earlier times in ancient civilization such as Egypt, China, Mesopotamia, and Mesoamerica (Eskelson, 2020; Law et al., 2015; Qargha & Morris, 2023). However, modern formal schooling, as it is known and practiced today, which transits from compulsory primary school education beginning from a young age and subsequently to higher levels, can be said to have begun in Europe, specifically in Prussia (Goldin, 2016).

Europe's Industrial Revolution and Human Capital Development

A second factor that has positively affected human capital development in Europe is the first industrial revolution. The industrial revolution in

---

[1] Many of these laws were passed to encourage people to read the bible in their vernacular, and reading literacy was a basic requirement for one to partake in some church activities (Antikainen & Pitkänen, 2014).

Western Europe sparked off unprecedented economic growth and brought about an increasing demand for human capital. Unlike the agrarian-based economy before it, industrialization required more advanced skills. Animal and manual human labor gave way to machinery such as the steam engine, spinning jenny, coke melting, and rolling (Mohajan, 2019). The new machinery and required value chains around the newly emerged industries brought about improved means of communications, transportation, banking, and related services. Machinery, mass production, and factories began to replace handcrafted goods. Consequently, the skills required to function in an economy that was previously primarily agrarian based were a lot different and less advanced than the ones required in the industrialized economy. Conversely, just as different forms of human capital became required after the industrial revolution, the revolution itself was sparked by human capital that had been accumulated in various forms such as occupation-specific skills and other forms of knowledge (Mokyr, 2017; Nuvolari et al., 2018; de Pleijt, 2020; Squicciarini & Voigtländer, 2015). There is growing evidence that upper-tail knowledge or upper-tail human capital, the human capital of a small group of elites of intellectuals and craftsmen, was a driving factor of the industrial revolution (See Mokyr, 2017; Voigtländer & Squicciarini, 2014; Squicciarini & Voigtländer, 2015).[2] Although the industrial revolution is said to have begun in Britain, it soon spread to other parts of Europe. Whether human capital was a direct cause, or only a resultant consequence, of the industrial revolution, Europe's human capital formation benefitted immensely from the first industrial revolution.

Other historical factors apart from literacy laws, school reforms, and the first industrial revolution have played definite roles in human capital formation in continental Europe. In more recent times, the economic and political integration that

have been enabled by the European Union has also encouraged a joint development of the continent's human capital of some sort. Consequently, many countries in Europe have an "above global average" human capital development, and this is reflected across the main global human capital indexes. Therefore, several countries qualify to be selected as the case study from Europe. Useful lessons on human capital development can be gleaned from Switzerland, Ireland, Sweden, Norway, the Netherlands, Austria, Denmark, Germany, the United Kingdom, and other European countries. Consequently, it was particularly tasking to select one of these countries because many of these countries performed well across the global human capital indexes. While the order of rankings for each of these countries varies across the indexes, the top five countries, in no particular order, appear to be Finland, Switzerland, Ireland, Norway, and Sweden. However, the case of Finland is especially adept for gleaning human capital developmental lessons because although Finland was one of the last European countries to implement compulsory schooling reforms in the last century (Gathmann et al., 2012; Hippe, 2020), it has surpassed other countries in human capital development. In addition, Finland has ranked as the number one country globally on many of the global human capital indexes. Finland was the topmost ranked European country in World Bank's HCI in 2018 and 2020 and on WEF's GHCI in 2015 and 2016. It was also the second-best European country on the GHCI in 2013 and 2017 and generally continues to rank among the top countries from Europe across all other global human capital indexes.

Finland: A Holistic Human Capital Development System

With a gross domestic product (GDP) per capita of over $50,500 and a population currently above 5.5 million, Finland is a country in Northern Europe and part of the Scandinavian and Nordic countries (World Bank, 2022). Originally part of Sweden until 1809 and a Grand Duchy of Russia between 1809 and 1917, Finland has grown from

---

[2] This is an emerging literature stream whose propositions and findings have ample support from other research streams (see Acemoglu et al., 2011; Gennaioli et al., 2013; Hanushek & Kimko, 2000).

an agrarian economy in the nineteenth century, through the period of partaking in the first industrial revolution and the post-second world war years, to becoming one of the industrialized countries of today. Although a welfare state, Finland has high standards of social and living conditions that is typical of Nordic countries and a low poverty rate (Keskimäki et al., 2019). Before the severe economic recession that gripped Finland in the 1990s, exports and forestry constituted the largest part of the economy. After recovery from the recession, by 1999, electronics, electrical equipment, and information and communications technology (ICT) became the largest parts of Finland's economy constituting roughly one-third of the country's GDP (Antikainen & Pitkänen, 2014; Keskimäki et al., 2019). Traditional industries of forestry and metal industries now coexist alongside the newly emerged industries in electronics, software, and other areas of technology (Keskimäki et al., 2019). Along with a network of small and medium enterprises, the global telecommunications giant, Nokia, formed the core of Finland's thriving ICT sector (Antikainen & Pitkänen, 2014; Schienstock, 2007). Schienstock (2007) attributes Finland's thriving ICT sector to "high investments in research and development and close science-industry co-operation, techno-organisational modernisation, a highly educated workforce, and a focus on firm-centred innovation policy" (p. 100).

Finland became a member of the EU in 1995 and has become globally renowned for its quality and holistic formal education system. Notably in 2006, for example, the country was the best performing country on the Programme for International Student Assessment (PISA),[3] a global examination organized by the Organisation for Economic Co-operation and Development (OECD) (Aedo et al., 2017; Hancock, 2011; Pulkkinen & Rautopuro, 2022). Indeed, Finland's education system caught the world's attention

with this and subsequent performances on PISA. Finland continues to perform well on all subject areas tested on PISA.

Interestingly, Finland's education system does not adhere to some of what can be regarded as global norms in formal education principles such as the use of mandated standardized national tests in basic primary and secondary education (Aedo et al., 2017; Hancock, 2011; Pulkkinen & Rautopuro, 2022; Välimaa, 2021). For example, apart from the National Matriculation Examination taken at the end of upper secondary school, there are no standardized tests in Finland's comprehensive schools (Välimaa, 2021). Public schools in Finland have continued to retain their quality and are free of tuition at all levels, offering the same curriculum and standard of education as private schools. This is in sharp contrast to what obtains in many countries where there is usually a wide disparity in the quality standards between public and private schools. On the health component side of human capital, Finland is also among the best performing countries in the world. A child born in Finland in 2022 has a 100% chance of surviving until age 5, and about 93% of children under 15 will live until they are 60 years old, and life expectancy at birth is 82 years (World Bank Human Capital Country Brief, 2022; World Bank, 2022). These statistics are reflections of a country with a healthy population that results from having quality health-care and education systems, in other words, quality human capital system.

School Reforms and Formal Education System in Finland

To understand the Finnish system of education, it is particularly noteworthy to look at it through a historical lens because Finland's current education system is a product of a series of reforms. Since undergoing its first major education reform in 1921, when the first compulsory schooling was introduced, the country has undergone other major schooling reforms. These major reforms in the education system that have culminated into the current system can be traced back to the

---

[3]PISA measures the ability of 15-year-olds in reading, mathematics, and sciences and their ability to use knowledge in these areas to meet life's challenges (OECD, 2023; Pulkkinen & Rautopuro, 2022).

1960s and 1970s (Antikainen & Pitkänen, 2014; Eklöf, 2018). Reforms into what became known as the comprehensive education system began after extensive negotiations with teachers' trade unions and other labor organizations and extensive preparation by various committees. The reforms did not result from hasty preparations and pronouncements. They were also not through a top-down approach in which government and their agencies solely make public policy decisions. Successful implementation of the many years of reforms has culminated in Finland being regarded by some as the country with the best education system in the world: at least in basic education which consists of primary and secondary school levels.

The Compulsory Education Act of 1921 made it compulsory for all Finnish children to complete 6 years of elementary schooling from age 7 (Gathmann et al., 2012; Hippe, 2020; Mikkonen, 2014). By this Act, compulsory schooling consisted of completing 6 years of primary school in six grades (Statistics Finland, 2007). This particular reform was so successful that in the mid-1930s, about 90% of Finnish children between 7 and 15 years were in school (Mikkonen, 2014; Statistics Finland, 2007). Prior to the enactment of the Act, only about 1% of the Finnish population above 15 years in age was literate (Antikainen & Pitkänen, 2014). In 1968, the school system Act was approved after careful and elaborate deliberations and preparations by various stakeholders (Aho et al., 2006; Antikainen & Pitkänen, 2014). This engagement with "state committees, negotiations with the teachers' trade organisations and other labour organisations, and experimentation by regional and local governments" (Antikainen & Pitkänen, 2014) was substantially essential to the success of the reform that continued to be implemented into the 1970s (Aho et al., 2006). This process elongated the enactment and eventual implementation of the reforms that have culminated into the current education system.

Finland's formal education system is sometimes described as a "miracle" due to the country's performance in the PISA tests and other global assessments such as Trends in International Mathematics and Science Study (TIMSS) and the Progress in International Reading Literacy (PIRLS) (Franco, 2020). Its education system since the reforms that began in 1921 into the 1970s and beyond has been underpinned by the philosophy of providing equal opportunities to a high-quality education for all Finns regardless of age, gender, or economic status (Aho et al., 2006; Eklöf, 2018; Leijola, 2004; Pulkkinen & Rautopuro, 2022). Education in Finland is seen as a basic human right that all Finns should have equal opportunities to acquire and not a privilege for the few. Consequently, all schools in the country are free. Both public and private schools are publicly funded. School meals are also free at the basic and secondary level and subsidized at higher education levels (Aho et al., 2006; Tuijnman & Hellström, 2001). This is an ample reflection of the welfare model of the Nordic countries, which Finland is part of, that advocates for equal social rights of citizens, public responsibility for the welfare of all citizens, income and gender equality, and full employment (Antikainen & Pitkänen, 2014). Finland's education system reflects a fundamental principle of the Nordic welfare model which is equal opportunities for all. To this end, the distribution network of schools is one that allows for schools to be close to homes and, when this is not possible, for example in rural areas, to provide a free and effective means of transportation to schools (Antikainen & Pitkänen, 2014; Lie et al., 2003).

A unique aspect of the Finnish education system is its flexible special education format that not only ensures equity but inclusion (Aedo et al., 2017; Välimaa, 2021). Each learner receives unique support as teachers differentiate their teaching to meet the needs of each student (Aedo et al., 2017). As a result, teachers may have to dedicate more time to a particular learner that needs more support than others. This support may be provided by the regular teacher or by the special education teacher in the form of regular support or intensified part-time inside or outside of the classroom (Aedo et al., 2017).

The current Finnish education system consists of compulsory basic education in comprehensive schools consisting of pre-primary, primary, and lower secondary schools (Education in Finland,

2022; Leijola, 2004). Children begin the comprehensive schooling system at 6 and primary schools at 7 and spend 6 years in primary school and 3 years in lower secondary schools as part of the compulsory comprehensive education structure (Education in Finland, 2022; Leijola, 2004). Post-basic school education system in Finland adopts the "duaali malli" or dual system comprising of upper secondary schools and vocation schools (Eklöf, 2018; Hancock, 2011; Leijola, 2004). Students from vocational secondary schools have access to tertiary education and are not limited in their choice of whether to continue along the vocation or university paths. Vocational education is competitive and develops into a higher education path. Students can advance into higher education either through the general or vocational path without any discrimination between both paths. Higher education consists of universities and polytechnics with polytechnics usually priding themselves with their connections with work and practice (Leijola, 2004).

A major backbone of the Finnish education system is the quality of its teachers. Research has established that merely completing formal education and measuring enrolment and completion rates are not the major factors of human capital formation. The quality of teachers and their teaching are one of the main determining factors of both the quantity and quality of human capital formation in individuals and in organizations[4] (Fomba et al., 2023; Tikly, 2019). The case of Finland is a worthy example of these research insights on teachers and the invaluable role they play in formal education and human capital formation. In Finland, teachers are selected from the top 10% of students and are required to earn a master's degree in education from a 5-year master's degree program (Aedo et al., 2017; Fornaciari & Juutilainen, 2023; Hancock, 2011). The master's program for teachers forms part of the country's university education system and has been required for entry into the teaching profession since 1978 (Aho et al., 2006). Teachers are

also required to undergo a Continuous Professional Development (CPD) annually (Education in Finland, 2022). Teachers are so valued in Finnish society that only a mere 10% of those that apply to become teachers are actually successful with their application to become teachers. Fornaciari and Juutilainen (2023) state that "teachers are highly respected professionals in Finnish society" (p. 1). A survey of teachers in Finland validated this statement as most teachers felt respected by the parents, school administration, and the general community (Popa et al., 2015). It suffices to say that the teaching profession is one that many in Finland will sure aspire to get into. This is indeed contrary to what obtains in many other countries where the teaching profession is seen only as a last resort for many, and sometimes, almost disdained as a career path.

Governance in Finland's formal education system is decentralized with local authorities, institutions, as well as teachers having significant amounts of autonomy. Teachers and education providers, for example, have the freedom to draw up their own curricula within the framework provided by the national core curriculum (Education in Finland, 2022). There are merits and demerits to having a decentralized governing structure. Given Finland's global reputation for quality education, it is safe to implicitly assume that such a decentralized governance system has been successful. However, Finland's quality education system cannot be said to be the result of any single feature. Well-implemented reforms, quality teaching and training standards, and other practices that are largely unusual to other climes have all played significant roles. Still, it is the integration of the features and the underlying principles derived from the fundamental Nordic philosophy that seem to undergird the success of the education system. The country is also not resting on its oars. Ongoing reforms continue in Finland with new policies and practices. Programs such as the Teacher Education Development Programme (TEDP), Action Plan for the New Comprehensive School, vision 2030 for higher education and research, and the right to learn programs all aim to continuously improve the education system

---

[4] "Organizations" is used here in the broad sense to refer to a collection or groups of individuals; this includes countries and societies.

with the special goals of improving equality and quality (OECD, 2020).

Adult Education

Human capital development in Finland goes beyond the formal education system that caters mainly to children and youths. Adult education is designed to provide opportunities for study for anyone between 18 and 64 years (Eurydice, 2022, 2023). This age range corresponds with the globally recognized working age group. Therefore, adult education in Finland is open to anyone within the working age group. Adult education, which takes place both formally and informally, is an important aspect of human capital development in Finland and the education system in Nordic countries (Eurydice, 2023; Tuijnman & Hellström, 2001). Informal adult education takes place in open colleges, workers' institutes, and other organizations, while universities, vocational schools, and other private institutions such as adult education centers provide formal adult education (Leijola, 2004). The major differences between informal and formal adult education are the certificate-oriented trainings of formal adult education and the government regulation of the curricular used by formal providers of adult education (Eurydice, 2022; Leijola, 2004). Adult education aims to strengthen the competence of the workforce, raise the employment rate by increasing the employability of adults, improve productivity, and generally encourage lifelong learning (Eurydice, 2022). An active adult education system serves to promote continuous human capital development. Adults are able to update their competences and human capital levels to current standards even if they had attended formal education in their younger years. The opportunity to change occupation, especially through attending adult vocational schools, is also enhanced (Eurydice, 2022; Leijola, 2004). About 1.1 million adults take part in adult education annually in Finland, and most courses are free with the exception of specialized courses (Eurydice, 2022; OPH, n.d.).

Health-Care System in Finland

Health care in Finland is provided by public and private health-care service providers. It is categorized into primary health care and specialist health care. Primary health care is provided mainly in health-care centers, while specialist care is provided by hospitals and private health-care providers. All residents are entitled to available health care in public and private hospitals. Similar to the education system in Finland, the health system was until recently a highly decentralized health-care system. In the highly decentralized system, although legislation and general policy guidelines were provided at the national level, local authorities known as municipalities were responsible for organizing primary and specialized health care (Keskimäki et al., 2019). However, the process of reducing the decentralization of health care in a major reform, which started with debates and different proposals in the 2000s, has been largely completed in 2023. From 2023 and beyond, in the new structure, health care is organized by 22 Well-Being Service Counties (WBSC) governed by democratically elected councils (European Observatory, 2023; Tynkkynen et al., 2023). The new system is expected to reduce inefficiencies that resulted from about 300 municipalities that administered the old outgoing health-care system. The WBSCs, along with the central government, will now be responsible for funding health care based on a formula (Tynkkynen et al., 2023). Health-care indexes in Finland are generally above European and global averages with good rankings on the Healthcare Access and Quality Index (HAQ): a testament to the effectiveness of the health-care system (Fullman et al., 2018; Tynkkynen et al., 2023). The restructured system is aimed at further improving the system to reduce existing inefficiencies and socio-economic and geographical inequalities; ensure the quality of health, social, and rescue services; improve access to care, particularly primary care; and control costs (Tynkkynen et al., 2023).

Human Capital Development Lessons from Finland, Nordic Countries, and Europe

The first lesson is the need for African countries to have compulsory schooling laws to promote reading, numeracy, and basic digital literacy. It is surprising that basic literacy and numeracy, which can be acquired through primary education, is still largely missing in some countries. Millions of children still lack access to primary school education in many African countries. Recent data from UNESCO shows that about 250 million children between the ages of 6 and 18 are out of school (UNESCO, 2023). Passing a compulsory schooling law is often not sufficient. The strategy and policies put in place for implementation and ensuring adherence to the law are often as important as dictates of the passed law. It's no surprise that some countries with compulsory schooling regulations still have challenges with getting all children to go to school. In an era of advanced digital technologies, inculcating digital literacy into compulsory schooling in addition to numeracy and reading literacy clearly becomes a necessity. Consequently, African countries will need to not only pass compulsory schooling laws but also put structures in place to ensure implementation.

A second subtle lesson from Europe is what can be regarded as a continental bandwagon effect—the compulsory school reforms in Europe seemed to have swept across the continent like a bandwagon. Countries implemented similar compulsory school reforms that they had seen in other surrounding countries. European countries have also benefitted from EU's cohesion policies that aim for regional development and reduction of disparities among member countries. There are various human capital policies in the integrated political and economic bloc that serve to increase human capital formation and development. Member countries have financial and other forms of support for human capital development from EU, and the regional policies also inform national policies and reforms. Although the world has increasingly gone global and the African continent is making efforts at better integration, there

seems to have been little bandwagon effect among African countries. The reasons for this situation are arguable. However, a human capital development policy or strategy with clearly visible positive national outcomes in one African country could easily become amenable for replication in another African country.

With a unique welfare economic model that is simultaneously able to aid a competitive private sector and create favorable business environment amid a largely viable economy, the Nordic economic system is sometimes regarded as a paradox. Education and health care, the two core components of human capital development, are largely free in Nordic countries. The free education system has been used by Nordic countries as a tool for providing equal opportunities for all (Deloitte Insights, 2020). However, this free human capital development system rests on values and philosophies that are inherently core to Nordic societies. These values have arisen partly from their unique historical contexts. Consequently, the lesson, and recommendation, is not the adoption of a free human capital development system. Rather, the lesson is that countries should develop their human capital in cognizance of their historical contexts and how these fit in with their envisaged vision. The path to progress in human capital development, and national development, rests on finding the path that builds on the past while building the "walk" toward the future. It will be difficult, for example, for countries with poor taxation and tax collection systems, which do not have similar values and generally lack the financial resources, to adopt a free human capital development system exactly like the Nordic system. However, some levels of compulsory education need to be free to encourage compliance with the compulsion. Indeed, compulsory formal school education must be accompanied by free education; otherwise, those without the means to pay for it become inherently disadvantaged by any compulsory schooling policies.

The case of Finland highlights the invaluable importance of stakeholder engagement in human capital developmental reforms and human capital development generally. A top-down approach can rarely provide the needed answers and solutions. Compared to similar developed countries, Finland's expenditure on education per-pupil before it earned global recognition for its education prowess through PISA can be regarded as meager (Aho et al., 2006). Therefore, answers and solutions to challenges in human capital development do not lie at merely increasing education and health-care expenditures. Careful thinking and preparation are also required. One of the best ways to effectively achieve this is through engagement with all relevant stakeholders, especially those that will be at the core of implementation and end users. Consequently, as in the case of Finland, reforms in education and health care cannot afford to be conducted in a haphazard manner and using a top-down approach. African countries that embark on education reforms will need to engage all stakeholders, starting at the grassroot and community levels in both urban and rural areas.

Teacher training has also been crucial to the quality prowess of the Finnish education system. The teaching profession is a unique profession that should be taken seriously. It is a profession that has a huge impact not only directly on the recipients of the services it offers but also indirectly on the society and country at large and on future generations. The case of Finland highlights the importance of professionalizing the teaching profession and ensuring that teachers meet the highest standards in academic excellence and necessary qualifications: educational and noneducational qualifications. Studies conducted by the Organisation for Economic Co-operation and Development (OECD) show that in Finland, teachers feel valued in society and enjoy competitive salaries comparable to other tertiary-educated adults in other professions (OECD, 2020). Consequently, the teaching profession should not be one that is perceived as being for those who have no other career option, as it is currently in many African countries. The teaching profession should be one that people aspire to. Finland, and indeed the Nordic countries, has shown that societies can progress as one whole when equal access to education and opportunities is provided.

The success of the Finnish education system cannot be disconnected from its historical origins and the fundamental beliefs and philosophies of the Nordic welfare system. Each Nordic country differs in the specific implementations of the fundamental beliefs and policies. However, the philosophies remain the core that holds the different implementations and manifestations in various economic and social spheres. Consequently, merely adopting one or some of the concepts of the Finnish education system might not be enough to replicate the success of the Finnish education system. The onus is on having a holistic education system that fits together and that is able to meet the present and future needs of the society and country. The education system should of necessity be one that aligns with the country's strategic vision and one that is capable of driving the country toward the vision. Since the education reforms that began majorly in the 1960s, education reforms in Finland continue and are ongoing with the latest reform conducted in 2019 and implemented between 2019 and 2023. Finally, human capital development should be a basic human right that every human being should have access to.

Conclusion

Historically, Europe has shown that countries sharing the same continent can and do indeed learn from one another. In modern economies, education systems are an integral part of the society and core to the human capital development system of any country. Finland's education system has shown how a quality holistic system of education can be developed. While it is not necessarily advisable for countries to wholly adopt the Finnish education, and health-care, systems, there are definitely lessons to be gleaned from them for countries that are looking for ways to enhance their human capital development system.

References

Acemoglu, D., Hassan, T. A., & Robinson, J. A. (2011). Social structure and development: A legacy of the holocaust in Russia. *Quarterly Journal of Economics, 126*(2), 895–946.

Aedo, C., Alasuutari, H., & Valijarvi, J. (2017). *Finland: A miracle of education?* A World Bank Blog. Available at https://blogs.worldbank.org/education/finland-miracle-education

Aho, E., Pitkänen, K., & Sahlberg, P. (2006). *Policy development and reform principles of basic and secondary education in Finland since 1968.* World Bank Education working paper series 2.

Antikainen, A., & Pitkänen, A. (2014). A history of education reforms in Finland. In R. R. Verdugo (Ed.), *Educational reform in Europe: History, culture, and ideology.* Information Age Publishing.

Becker, S. O., & Woessmann, L. (2009). Was Weber wrong? A human capital theory of protestant economic history. *The Quarterly Journal of Economics, 124*(2), 531–596.

De Pleijt, A. M. (2020). A tale of two "educational revolutions". Human capital formation in England in the long run. *Dans Revue D'Economie Politique, 130,* 107–130.

Deloitte Insights. (2020). *The Nordic social welfare model: Lessons for reform.* A publication of Deloitte Development LLC, Deloitte Touche Tohmatsu Limited.

Education In Finland. (2022). *Finnish education in a nutshell.* A publication of the Ministry of Education and Culture Finland, and Finnish National Agency for Education, 2022 series.

Eklöf, T. (2018). *How have changes in human capital investment impacted the development of the national economy of Finland?* [Bachelor's Thesis, Helsinki Metropolia University of Applied Sciences, Finland].

Eskelson, T. C. (2020). How and why formal education originated in the emergence of civilization. *Journal of Education and Learning, 9*(2), 29–47.

European Observatory. (2023). *Finland: Health summary.* Available at https://eurohealthobservatory.who.int/publications/i/finland-health-system-summary

Eurydice. (2022). *Adult education and training.* Available at https://eurydice.eacea.ec.europa.eu/national-education-systems/finland/adult-education-and-training

Eurydice. (2023). *Adult education and training.* Available at https://eurydice.eacea.ec.europa.eu/national-education-systems/finland/adult-education-and-training

Fomba, B. K., Talla, D. F., & Ningaye, P. (2023). Institutional quality and education quality in developing countries: Effects and transmission channels. *Journal of the Knowledge Economy, 14,* 86–115. https://doi.org/10.1007/s13132-021-00869-9

Fornaciari, A., & Juutilainen, M. (2023). Who has the power to define the ideal teacher? Insights into the social structure of Finnish teacher education. *Frontiers in Education, 8,* 1297055.

Franco, A. (2020). Not all Finns think alike: Varying views of assessment in Finland. *International Education Studies, 13*(1), 1–10. https://doi.org/10.5539/ies.v13n1p1

Fullman, N., et al. (2018). Measuring performance on the Healthcare Access and Quality Index for 195 countries and territories and selected subnational locations: A systematic analysis from the Global Burden of Disease Study 2016. *Lancet, 391*(10136), 2236–2271.

Gathmann, C., Jürges, H., & Reinhold, S. (2012). *Compulsory schooling reforms, education and mortality in twentieth century Europe,* IZA discussion papers, no. 6403. Institute for the Study of Labor (IZA). https://nbn-resolving.de/urn:nbn:de:101:1-201206146887

Gennaioli, N., La Porta, R., Lopez-de-Silanes, F., & Shleifer, A. (2013). Human capital and regional development. *Quarterly Journal of Economics, 128*(1), 105–164.

Goldin, C. (2016). Human capital. In C. Diebolt & M. Haupert (Eds.), *Handbook of cliometrics.* Springer Verlag.

Hancock, L. (2011). *Why are Finland's schools successful?* Available at https://www.smithsonianmag.com/innovation/why-are-finlands-schools-successful-49859555/

Hanushek, E. A., & Kimko, D. D. (2000). Schooling, labor-force quality, and the growth of nations. *The American Economic Review, 90*(5), 1184–1208.

Hippe, R. (2020). Human capital in European regions since the French revolution: Lessons for economic and education policies. *Dans Revue D'Economie Politique, 130,* 27–50.

Keskimäki, I., Tynkkynen, L. K., Reissell, E., Koivusalo, M., Syrjä, V., Vuorenkoski, L., Rechel, B., & Karanikolos, M. (2019). Finland: Health system review. *Health Systems in Transition, 21*(2), 1–166.

Law, D., Wang, H. C., Nissen, H. J., & Urton, G. (2015). Writing and record-keeping in early cities. In Y. Norman (Ed.), *The Cambridge world history* (Vol. III: Early cities in comparative perspective, 4000BCE–1200CE). Cambridge University Press. https://doi.org/10.1017/CHO9781139035606.013

Leijola, L. (2004). *The education system in Finland: Development and equality.* ETLA, Elinkeinoelämän Tutkimuslaitos, The Research Institute of the Finnish Economy, (Keskusteluaiheita, discussion papers, ISSN 0781-6847; no. 909).

Lie, S., Linnakylä, P., & Roe, A. (2003). *Northern Lights on PISA: Unity and diversity in the Nordic countries in PISA 2000.* Department of Teacher Education and School Development, University of Oslo.

Mikkonen, J. (2014). *Educational policies: Finland.* Available at https://splash-db.eu/policydescription/educational-policies-finland-2014/#:~:text=Under%20the%20Compulsory%20Education%20Act,smallest%20and%20most%20remote%20municipalities

Mohajan, H. K. (2019). The first industrial revolution: Creation of a new global human era. *Journal of Social Sciences and Humanities, 5*(4), 377–387.

Mokyr, J. (2017). Bottom-up or top-down? The origins of the industrial revolution. *Journal of Institutional Economics, 14*(6), 1003–1024.

Nuvolari, A., Weisdoff, J., & de Pleijt, A. (2018). *Human capital formation during the first industrial revolution: Evidence from the use of steam engines.* VoxEU. https://cepr.org/voxeu/columns/human-capital-formation-during-first-industrial-revolution-evidence-use-steam-engines

OECD. (2020). *Education policy outlook: Finland.* Available at https://www.oecd.org/education/policy-outlook/country-profile-Finland-2020.pdf

OECD. (2023). *What is PISA?* Available at https://www.oecd.org/pisa/

OPH. (n.d.). *Liberal adult education.* Finnish Agency for Education. Available at https://www.oph.fi/en/education-system/liberal-adult-education

Popa, C., Laurian, S., & Fitzgerald, C. (2015). An insight perspective of Finland's education system. *Procedia – Social and Behavioral Sciences, 180,* 104–112. https://doi.org/10.1016/j.sbspro.2015.02.092

Pulkkinen, J., & Rautopuro, J. (2022). The correspondence between PISA performance and school achievement in Finland. *International Journal of Educational Research, 114,* 102000. https://doi.org/10.1016/j.ijer.2022.102000

Qargha, G. O., & Morris, E. M. (2023). *Why understanding the historical purposes of modern schooling matters today. A bookings commentary.* Available at https://www.brookings.edu/articles/why-understanding-the-historical-purposes-of-modern-schooling-matters-today/

Schienstock, G. (2007). From path dependency to path creation: Finland on its way to the knowledge based economy. *Current Sociology, 55*(1), 92–109.

Squicciarini, M. P., & Voigtländer, N. (2015). Human capital and industrialization: Evidence from the age of enlightenment. *The Quarterly Journal of Economics, 130*(4), 1825–1883.

Statistics Finland. (2007). *Education in Finland: More education for more people.* Available at https://www.stat.fi/tup/suomi90/marraskuu_en.html

Tikly, L. (2019). Education for sustainable development in Africa: A critique of regional agendas. *Asia Pacific Education Review, 20,* 223–237. https://doi.org/10.1007/s12564-019-09600-5

Tuijnman, A., & Hellström, Z. (Eds.). (2001). *Curious minds: Nordic adult education compared.* Nordic Council.

Tynkkynen, L. K., Keskimäki, I., Karanikolos, M., & Litvinova, Y. (2023). *Finland: Health system summary, 2023.*

UNESCO. (2023). *Out-of-School rate.* Available at https://education-estimates.org/out-of-school/

Välimaa, J. (2021). Trust in Finnish education: A historical perspective. *European Education, 53*(3–4), 168–180. https://doi.org/10.1080/10564934.2022.2080563

Voigtländer, N., & Squicciarini, M. (2014). *Knowledge elites, enlightenment, and industrialization,* VoxEU. https://cepr.org/voxeu/columns/knowledge-elites-enlightenment-and-industrialisation

World Bank. (2018). *The human capital project.* World Bank Report 2018.

World Bank. (2022). *Finland: Overview.* Available at https://data.worldbank.org/country/FI

World Bank Human Capital Country Brief. (2022). *Finland.*

Immigration and Human Capital Accumulation in North America: The Case of Canada

8

Abstract

This chapter presents Canada's distinctive approach to immigration that has formed one of the cores of the country's human capital accumulation and development strategy. From a historical perspective, international immigration has always formed an important source of human capital acquisition in Canada. This aligns with the country's official adoption of multiculturalism which is also embedded in the country's Charter of Rights and Freedom. This chapter traces immigration in Canada over the years and presents an overview of some of the major policies and programs that have been used to encourage immigrants to come in and integrate into the Canadian society and economy. This chapter concludes with lessons from Canada's approach to human capital development and accumulation using immigration.

Keywords

Strategic immigration · International immigration · Immigration policy · Human capital · Canada

Introduction

The continent of North America is one that consists of both sovereign countries and non-sovereign territories. Among the 23 sovereign countries in North America, Canada, the United States (USA), and Mexico stand out as the top three countries across different global human capital indexes. As the only continent that is grouped alone on most of the global human capital indexes, it is easily ranked as the top performing content. On the World Banks's Human Capital Index (WB HCI) ranking of 2020, for example, the continent's overall index was 0.75 with the grouping of Europe and Central Asia coming a distant second with 0.69. For comparison, that of Sub-Saharan Africa (SSA) was 0.40, South Asia was 0.48, Latin America and Caribbean was 0.56, and Middle East and North Africa was 0.57 on the index. East Asia and Pacific completed the continental groupings and had index points of 0.59.

However, this noteworthy group performance of North America masks the lackluster performances of many of the countries on the continent. The overall performance of the continent is greatly boosted by the performance of the top three countries of Canada, the USA, and Mexico. The Global Talent Competitiveness Index (GTCI), published by Insead, Human Capital Leadership Institute, and Portulans Institute ranked only Canada and the USA as North

American countries in 2022, ranking Mexico as a Latin American country. On the index, North America, especially named Northern America, is the top performing continental groupings of all the groupings. Other human capital indexes also routinely have different continental groupings that give the impression of North America's top performance on the indexes because of the top ranking positions of Canada, the USA, and in some cases, Mexico, depending on the continental groupings. Undoubtedly, Canada and the USA have made tremendous progress in developing and utilizing human capital. However, Canada and the USA, unlike other countries in North America, can be regarded as countries of immigrants because international immigrants constitute a major portion of their populations. These two countries were also historically "established" by immigrants primarily from Europe and other continents.

Therefore, although this chapter on immigration as a source of human capital acquisition focuses on Canada as its case study, it is case worthy to mention some important facts about human capital development in the USA. Early historical development of human capital in the USA was spearheaded by those that were largely generations of the founding immigrants into the country. Mass primary and secondary school education began in the USA in the nineteenth century, and available data show that the USA was far ahead of Europe in this regard at this time (Goldin, 2001, 2016). The mass education at formal secondary school education in the early parts of the twentieth-century USA, which were still largely unavailable in Europe, has been argued to account for much of the difference in the economic growth and development between the USA and European countries (Goldin, 2001, 2016) and, indeed, the rest of the modern world—available data from Barro and Lee (2015) and Lee and Lee (2016) on average years of schooling show that North America, as a continent, was far ahead of other continents at this time in history, lending additional credence to this argument. Research on modern economic growth models such as endogenous growth model show that inequalities in economic growth and devel-

opment among nations are largely due to differences in their human capital development and accumulation (Galor & Weil, 2000; Mankiw et al., 1992; Sharipov, 2015). Therefore, it is not very surprising that across all global human capital indexes, the rankings somewhat reflect the level of countries' growth and development. There is a strong positive correlation between countries that are highly ranked on human capital indexes and their level of development. Industrially advanced countries have developed and accumulated human capital over time much more than developing countries. However, countries have approached human capital development and accumulation in several different ways often using multiple means and approaches.

Canada's Approach to Human Capital Accumulation and Development

The country of Canada is a vast geographical land that extends across six time zones and eight climatic regions (Martin et al., 2018). With a gross domestic product (GDP) per capita of US$54,966 (World Bank, 2022a) and a population of about 39 million (Statistics Canada, 2023; World Bank, 2022b), Canada is the second largest country in the world by geography. It is one of the 56 independent countries that are members of the Commonwealth having resulted from a merger of 3 British colonies which constituted the initial provinces of the country in 1867. Canada ranks highly on many of the global indexes for economic advancement, innovation, and competition and on human capital development. The country has an impressive pool of people with highly skilled human capital (Insead, 2022; OECD, 2019). This is as a result of a multipronged approach to human capital development that emphasizes education and skills. The country ranked 5th on Insead's GTCI of 2022 and on World Bank's HCI of 2022 and 15th on United Nations' Human Development Index (HDI) for 2021 and 2022. It has consistently ranked as the number one country in North America, geographically, in many of the available global human capital indexes. Clearly, Canada has some human capital policies and programs in place that are worth studying. There are lessons to be gleaned even from a mere cursory study of human capital development and accumulation in the North American country.

Canada's approach to human capital has been multipronged through its education and training system and notably through immigration of highly skilled workers. This multipronged approach has largely resulted from deliberate policies aimed at addressing workforce shortages due to an aging population, emigration of skilled workers to the USA, declining dependency ratio, and low and slowing birth rates. In the decades between 1977 and 2017, the number of people who were 65 years and older in Canada rose from 2 million to 6.2 million and is expected to reach 10.4 million by 2037 (CIHI, 2017). In 2022, 19% of Canada's population were 65 years or older (Fraser Institute, 2022). The number of births per a thousand people has gone down from 26.8 to 9.8 in 2023 (UN, 2023). Since 1999, increases to Canada's population have resulted majorly from international immigration rather than from child birth or natural increase, which is the net of child births and deaths in the population (Statistics Canada, 2018, 2023). In 2022, a record increase in population, which is Canada's second highest annual increase in population growth rate, resulted from international immigration (Statistics Canada, 2023).

An aging population coupled with low and slowing birth rate can have potential challenges for the workforce, as well as for public finance, welfare, and utilities, of any country. Favorable policies that encourage international immigration into Canada from other countries have been a major throng of the country's human capital development policies. Between 2016 and 2021, 80% of the growth in Canadian workforce resulted from immigration (Statistics Canada, 2022a). For a country that has historically been populated largely by immigrants, the current high rate of immigration seems to be some form of conformity with its historical context and antecedents.

A Brief History of International Immigration in Canada

Canada has a long history of international immigration. In fact, Canada can be aptly regarded as a country of immigrants. The indigenous peoples of India, Inuit, and Métis were living in modern-day Canada before the arrival of the European immigrants who settled on the land in the sixteenth and seventeenth centuries (Britannica, n.d.; Government of Canada, 2018; University of Ottawa, n.d.). However, since the confederation of the first three provinces of Canada in 1867, over 17 million immigrants have come into Canada to settle (Martin et al., 2018; Statistics Canada, 2016), displacing the indigenous population. This figure has subsequently increased since 2016, and immigrants represent about 23% of the Canadian population (Statistics Canada, 2022a, b). Many of the first immigrants into Canada came notably from Britain and France into different regions, which were then separate and later became British colonies before the confederation in 1867.

The first law enacted to govern immigration in Canada was in 1869, just 2 years after the official merging of the colonies into a single country. The first Immigration Act of 1869 was aimed at encouraging immigration and ensuring the safety and protection of immigrants upon their arrival into the relatively newly formed country (Van Dyk, 2023). Most of the immigrants subsequent to this immigration act came primarily from Europe (Statistics Canada, 2022a). The Immigration Act of 1869 served as an open and nondiscriminatory policy to immigration. Between 1896 and 1915, about 3 million immigrants came into Canada from the USA, Britain, and other European countries due to an active drive to bring in immigrants to work in the Prairie agricultural farmlands and the then emerging manufacturing and railway construction sectors (Cheatham & Roy, 2023; History Museum, 2023; Simmons, 2010). The Canadian government used immigration to develop some of the vast land through different information campaigns often to many European countries to encourage immigration into Canada (Cheatham & Roy, 2023).

Between 1871 and 1931, Canada saw a surge of immigrant population, mostly from Europe (see Fig. 8.1).

Immigrants also came into Canada from Asia and the Middle East. However, the open-door approach to immigration that was heralded by the Immigration Act of 1869 soon gave way to one that quickly became overtly discriminatory until 1962 when many racial discrimination elements were eliminated from Canadian immigration policy (Fitzgerald & Cook-Martín, 2014; Lake & Reynolds, 2008; Triadafilopoulos, 2022; Van Dyk, 2023). Notable immigration acts that served to limit immigration from certain parts of the globe include the Chinese Immigration Act of 1885, which imposed payment, known as head tax, for Chinese persons entering into Canada, the immigration agreement with the Japanese government in 1908 that sought to limit the number of Japanese immigrants into Canada, and the Immigration Act of 1906 which empowered the government to prohibit the entrance of certain immigrants as it deemed necessary (History Museum, 2023; Van Dyk, 2023). The Chinese Immigration Act of 1923 also sought to limit and further restrict immigration of Chinese persons into Canada (Van Dyk, 2023). Consequently, by the end of the second world war in 1945, Canada was populated almost exclusively by immigrants from Europe (History Museum, 2023): what

some regard as an attempt to keep Canada a "white man's country" (Challinor, 2011; Fitzgerald & Cook-Martín, 2014; Triadafilopoulos, 2022).

Canada only officially adopted multiculturalism as a policy in 1971 to promote cultural diversity and to encourage international immigrants from other countries to come into Canada (Van Dyk, 2023). This policy was given a formal legislative framework by the Multiculturalism Act of 1988 and entrenched into the Canadian Charter of Rights and Freedom (Triadafilopoulos, 2021; Van Dyk, 2023). Prior to this official adoption of multiculturalism, many of the precedent immigration policies were clearly and largely discriminatory against non-European immigrants. However, a "merit-based" system based on points for evaluating immigrants was introduced in 1967 (Challinor, 2011; Cheatham & Roy, 2023; Triadafilopoulos, 2022). Criteria other than race became the standard for measuring applicants' eligibility for immigration into Canada. Prospective immigrants were assessed based on education level, training, knowledge of English and French, and other criteria related to employment history and prospective employment in Canada (Triadafilopoulos, 2022). Summarily, human capital, rather than race, became the primary criterion for international immigration assessment into Canada. Additional programs to

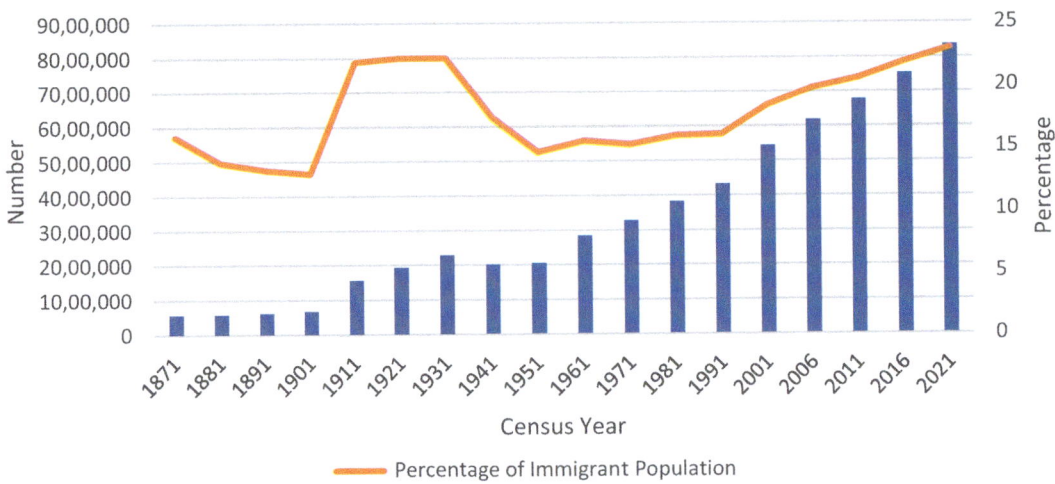

Fig. 8.1 Number and percentage of immigrant population in Canada, 1871–2021. (Source of data for graph: Statistics Canada 2022b)

attract entrepreneurs and investors were introduced in 1986 (Simmons, 2010; Triadafilopoulos, 2022).

Subsequent to the adoption of multiculturalism within a bilingual (English and French languages) framework, the Immigration Act of 1976 officially outlined the objectives and general planning process of the country's immigration policies (Cheatham & Roy, 2023; Triadafilopoulos, 2022; Van Dyk, 2023). The Immigration Act of 1976 set out to achieve the main goals of promoting economic, social, demographic, and cultural goals and family unification, and to ensure Canada fulfills its obligations under the United Nations convention and protocol on refugees (Triadafilopoulos, 2022). These three broad objectives are neatly fitted into three different classes of immigrants: economic, family, and refugees (Triadafilopoulos, 2022). Canada also admits immigrants based on humanitarian and compassionate grounds (Cheatham & Roy, 2023). Since the adoption of multiculturalism, the number of immigrants into Canada has seen a steady increase from just below 3.3 million people in 1971 to over 8.3 million people in 2021. From constituting about 15.3% of the country's population in 1971, immigrant population constituted 23% of the population in the last census of 2021 (see Fig. 8.1).

Contemporary Policies on International Immigration (2002 till Date)

Historically, it can be seen that Canada has always deliberately and strategically, in many instances, managed immigration while taking the economy into consideration. Canada has always sought to use immigration as a tool for economic growth and to replace emigration to other countries, especially emigration to the USA (Kaushik & Drolet, 2018; Thompson & Weinfeld, 1995; Triadafilopoulos, 2022). This general objective has not changed in contemporary times. A major goal of Canada's contemporary immigration system is to attract youthful and highly skilled immigrants with the requisite human capital to fill the

workforce. Certain specific education criteria, skills, experience, and language abilities in English and French languages are the fundamental requirements to gain entry into Canada (Challinor, 2011; Triadafilopoulos, 2022). Different immigration programs put in place aim to ensure only certain categories of immigrants are let into the country. Thus, although Canada's immigration system is still discriminatory, the basis of discrimination is no longer race. International immigrants continue to come into Canada either as economic migrants, family, refugees and protected persons, or on humanitarian grounds (Cheatham & Roy, 2023; Triadafilopoulos, 2022). Economic migrants are those with the required human capital to work or financial resources to invest and start businesses within Canada. Family immigrants are those who wish to join and settle with their family members that are already in Canada.

The points-based system introduced in 1967 has undergone several revisions and is complemented by the Immigration and Refugee Protection Act (IRPA) of 2001, which replaced the Immigration Act of 1976. The IRPA came into effect in 2002 and provides the broad framework that has guided subsequent reforms to Canada's international immigration system. The Act emphasized education, language, and adaptability as the general criteria for immigration (Challinor, 2011). The broad and scanty nature of the IRPA affords the executive government much liberty in setting immigration policies and programs. Consequently, different initiatives have been introduced and subsequently revised over the years in the bid to adjust immigration numbers to meet the country's social and economic realities. The Federal Skilled Worker Program (FSWP), introduced in the 1970s, for example, has undergone a series of changes and has been improved through the express entry system (Triadafilopoulos, 2022). FSWP is now part of the Federal High Skills along with the Federal Skilled Trades Program and Canadian Experience Class (Government of Canada, 2022a). All three routes under the Federal High Skills—the FSWP, Federal Skilled Trades Program, and the Canadian Experience Class (CEC)—are now managed using the express entry system, an online immigration management system for skilled workers (Government of Canada, 2023a). Other immigration routes through which economic immigrants can come into the country include the federal business route made up of start-up visa program and the Self-Employed Persons Program and Provincial Nominee Program (PNP) for different provinces that enable provinces to select economic immigrants to meet their demographic and labor market needs (Banting, 2010; Government of Canada, 2022a; Paquet, 2014; Triadafilopoulos, 2022). International students who study in Canada can also become permanent residents by going through the CEC route (Challinor, 2011).

The number of immigrants into Canada has continued to increase annually regardless of the political party in power (Triadafilopoulos, 2021). Since 2017, the Canadian government has set out multi-year immigration levels plan using a 3-year planning horizon (Government of Canada, 2022b). The government recently announced its plan to let in between 465,000 and 500,000 people between 2023 and 2025 through the four main classes of immigration: economic, family, refugee, and humanitarian (Government of Canada, 2022a, 2023c).

Human Capital Development and Integration for International Immigrants

There are peculiar challenges that result from high immigration. In addition to the challenges from having a multicultural society, there can be a strain on public services. Housing, transportation, and other infrastructure can become overstretched. Immigrants themselves may find it difficult to settle into the country and integrate into the society. There have been tales of unemployment and underemployment of immigrants with advanced degrees and professional designations on arrival in Canada (Boyd, 2013; Challinor, 2011; Hawthorne, 2013; Triadafilopoulos, 2022). Immigrants have sometimes been unable to find jobs commensurate with their level of education and experiences (Boyd, 2013; Challinor, 2011),

leading to underutilization of their human capital. However, integration goes beyond finding employment. Immigrants can only be said to have integrated when they can contribute to the economic, social, cultural, and political dimensions of the Canadian life (Kaushik & Drolet, 2018; Murphy, 2010). Working and having a sense of belonging in the community is important to self-worth of immigrants and for integration (Murphy, 2010). To mitigate some of these integration challenges, legal immigrants that have made use of the different immigration routes are offered orientation programs, skills training, and social services to enable them assimilate into the country (Cheatham & Roy, 2023). For example, non-English-speaking immigrants go through a government-funded language training upon arrival into the country for some weeks (Martin et al., 2018). Information and guidance on how to settle into Canada are also now provided to immigrants prior to their leaving their home countries (Kaushik & Drolet, 2018). Municipalities also provide support for initiatives that encourage immigrants to participate in the community and anti-racism (Banting, 2010). These actions are in a bid to encourage social cohesion and prevent the kind of social tensions that have taken place in some other countries that have attempted strategic immigration of skilled workers.

Strategic Use of Immigration as Human Capital Acquisition

The bouts of immigration in the nineteenth and early twentieth centuries served as labor input into a then emerging economy. Workers were needed in various sectors of the economy to grow the economy (Thompson & Weinfeld, 1995). Contemporary immigration policies have also served very similar purposes—the acquisition of human capital. Thus, immigration in Canada has always served as a means of taking in people primarily because of the human capital embodied in them and how much they can potentially contribute to the country's economic growth. The standards for immigration into Canada may have changed over the years from one that was overtly racially discriminatory to one that is currently more open to racial diversity. However, discrimination remains. In current times, discrimination is primarily based on human capital. This can be seen from the increasing number of economic immigrants over the years. Economic immigrants constitute more than half of recent immigrants into Canada (Challinor, 2011; Government of Canada, 2022a). Even in managing immigration refugees, there is concerted effort to ensure that human capital remains a key consideration for admission of refugees. The Economic Mobility Pathways Pilot (EMPP) program, for example, first launched in 2018 and relaunched in 2021, aims to bring in skilled refugees to meet the demands of the labor market (Government of Canada, 2022b). Other refugee programs in place such as the government-assisted and privately sponsored refugee resettlement programs are carefully monitored and controlled. Only few refugees come in through the country's asylum programs, and overall, immigrants that come in through refugee programs constitute the least of the three main classes of immigration (Triadafilopoulos, 2022). Hence, the focus remains human capital acquisition through immigration in order to ensure that immigrants contribute, and do not become a burden, to the economy.

Canada's Health-Care System

Human capital development in any country inherently includes the health-care system because good health is a fundamental component of human capital (Becker, 2007; Bloom & Canning, 2003). Canada operates a universal health publicly funded health-care system that offers free medical services to all its citizens and permanent residents. As a universal health-care system, its founding principle is health care that is based on need rather than on the ability to pay. The country's universal health-care system is funded through the "Medicare" system created by the Health Act of 1984 and provides free in- and out-patient care excluding the cost of prescription drugs. The journey toward the universal health-

care system began in 1966 with the passage of the Medical Care Act in 1966. However, the precursor to this national Act was the universal public health insurance of the Saskatchewan province (Government of Canada, 2019, 2023b; Martin et al., 2018). The Act made it possible for part of the medical expenses for services provided by doctors outside the hospital to be reimbursed by the government (Government of Canada, 2019). The health Act of 1984 improved on the many shortcomings of the Act of 1966. Since then, subsequent legislations and reforms have aimed at improving the Act and health-care system.

Medicare is not a single national health-care system but one that integrates the health-care systems in the ten provinces and three territories (Dhalla & Tepper, 2018; Martin et al., 2018). Consequently, although there are many similarities among the health-care system in the provinces and territories, the type and nature of health care also differs in some respects between the provinces and territories, despite being guided by common legislative frameworks which set the basic standards. Generally, the health Act of 1984 stipulates that provincial and territorial health-care systems must be portable, universal, accessible, comprehensive, and publicly administered (Martin et al., 2018; Government of Canada, 2019, 2023b). These five principal standards ensure that all eligible Canadian residents have uniform access, without any preferential access, to all publicly insured health-care services and are able to use the same health-care plan across the country (Martin et al., 2018). Responsibility for delivering of health care is shared between the federal, provincial, and municipality governments, under the regulatory guidance of the health act and other regulations. However, funding is primarily provided by the federal government. Eligible Canadians do not have to pay for health-care services as the services are free at the point of delivery; no reimbursement is required. To ensure everyone is covered, the health-care system also provides for the provision of some direct health-care services to specific groups of people such as the indigenous peoples living on reserves, serving members of Canadian forces, eligible veterans, federal prison inmates, and some groups of refugees (Government of Canada, 2023b).

However, there are reports of inadequacies in the current health-care system and calls for improvement (Dhalla & Tepper, 2018; Garrod et al., 2020; Martin et al., 2018; Patrick & Laupacis, 2023). Some eligible residents have reported difficulties in accessing some health-care services (Garrod et al., 2020; Patrick & Laupacis, 2023). Nevertheless, Canada, as a country, continues to improve and perform generally well on some national health indicators. Average life expectancy at birth, for example, has increased from 71 years in 1960 to 83 years in 2021 (World Bank, 2023). Canada also compares favorably well with other Organisation for Economic Co-operation and Development (OECD) countries on other national health indicators (Martin et al., 2018). Overall, despite the current challenges with the country's health-care system and the need for improvement, the system is one that provides some relative health-care support for human capital to strive.

Human Capital Development Lessons from Canada

Canada's historical and contemporary experience with human capital accumulation through immigration has been largely successful. Current evidence and research suggest that Canadian's experiments in immigration and multiculturalism have been successful so far (Thompson & Weinfeld, 1995; Triadafilopoulos, 2022). The Organisation for Economic Co-operation and Development (OECD) has applauded Canada's skilled immigration system as the most successful of such immigration system and a benchmark for other countries to emulate (OECD, 2019). Canada's immigration system has managed to expand from year to year while managing to gain broad acceptance within the country with few Canadian dissidents to the rising numbers of international immigrants into the country. Overall, the expanding immigration system enjoys broad consensus and support (Triadafilopoulos, 2022). Canada's immigration

system has been carefully designed to meet the needs of the country which has the highest number of highly educated immigrants among OECD countries (Cheatham & Roy, 2023; OECD, 2019) and perhaps in the world. The Canadian model of immigration has garnered so much global attention among scholars and researchers that a literature stream dedicated to its study has emerged (see Triadafilopoulos, 2021, for a number of publications on the subject).

A major lesson from Canada to African and other countries is that nontraditional sources such as immigration can be used to strategically acquire the desired human capital if and when necessary. Such programs should, however, be carefully designed to suit the requirements of the country and ensure that the desired goals and potential economic gains to the country are achieved without undue cultural, social, and economic hindrances to integration. There are some claims that the present success of Canada's immigration policy is due partly to the country's geographical isolation from other countries apart from the USA (Cheatham & Roy, 2023; Reitz, 2012; Triadafilopoulos, 2021). Arguably, this "natural luck" in geographical location has helped to limit influx of illegal immigrants who most likely would have otherwise come in en masse. However, the country's careful and effective design of immigration, coupled with an effective management of a comprehensive immigration system, deserves some commendation. The government's ability to receive broad support from the general public and from many opposition parties for its continuous immigration policies when other countries have failed to do so is also admirable. The mere fact that the general support levels have not waned despite the huge levels of current and planned immigration is evidence of the effective management of the immigration system, system of integration of immigrants, and management of perception.

In 1990, with a period human capital measure of 23 years, Canada was categorized as one of the top five (countries) globally by Lim et al.'s (2018) period measure of human capital. However, the country dropped to the 11th place, having a period human capital of 25 years in 2016. By this index, while there was an increase in absolute terms of 2 years to the period human capital of Canada between 1990 and 2016, there was a drop in the country's global ranking. This, perhaps, indicates the impact of recent immigration policies or a need for the country to assess its human capital development policies in totality rather than just a seeming focus on immigration as the route to increasing its stock of human capital. Between 1995 and 2020, immigrants accounted for 40% of the growth in human capital in Canada (Gu, 2023). However, Canada's performance on other global human capital indexes is still noteworthy. Perhaps, the case of Canada's strategic use of international immigration as a source of human capital accumulation showcases why this should not exclude other forms of human capital development.[1] It also highlights the fundamental principle in human capital that people, by extension, the population, form the very bedrock upon which sustainable development should be anchored. Human capital resides in people and cannot be alienated from them (Esho & Verhoef, 2020; Kim & Mahoney, 2007). Consequently, there is only so much a country with an aging population and low and reducing birth rate can do, especially in the short term, to increase and accumulate human capital.[2] Although African countries generally do not have an aging population or declining birth rate like Canada, strategic use of immigration to access the necessary human capital should not be overlooked as it has its usefulness.

Conclusion

North America, and specifically Canada, can be said to be a continent and country of immigrants. Right from the inception of the country, planned

immigration has served as a means of attracting people with the requisite human capital into the country and as a tool for economic growth. The case of Canada shows that when using immigration for human capital acquisition, countries must carefully design it to accommodate the socio-economic needs, and, very importantly, the geographical location of the country needs to be taken into consideration. In this regard, there is also need to ensure that the immigration strategy aligns with and takes cognizance of the political, economic, and social state and nature of neighboring countries.

References

Banting, K. (2010, November). *Federalism and immigrant integration in Canada.* Unpublished paper for the conference on Immigrant Integration: The Impact of Federalism on Public Policy. Brussels, Belgium.

Barro & Lee (2015). Lee & Lee (2016). With major processing by Our World in Data. "Average years of schooling" [dataset]. Barro and Lee, "Projections of Educational Attainment"; Lee and Lee, "Human Capital in the Long Run" [original data]. Retrieved August 30, 2024 from https://ourworldindata.org/grapher/mean-years-of-schooling-long-run

Becker, G. S. (2007). Health as human capital: Synthesis and extensions. *Oxford Economic Papers, 59,* 379–410.

Bloom, D., & Canning, D. (2003). Health as human capital and its impact on economic performance. *The Geneva Papers on Risk and Insurance, 28*(2), 304–315.

Boyd, M. (2013). Accreditation and the labour market integration of internationally trained engineers and physicians in Canada. In T. Triadafilopoulos (Ed.), *Wanted and welcome? Immigrants and minorities, politics and policy* (pp. 165–198). Springer.

Britannica. (n.d.). *Indigenous peoples.* Available at https://www.britannica.com/place/Canada/Government-and-society

Challinor, A. E. (2011). *Canada's immigration policy: A focus on human capital.* Migration Policy Institute. Available at https://www.migrationpolicy.org/article/canadas-immigration-policy-focus-human-capital

Cheatham, A., & Roy, D. (2023). *What is Canada's immigration policy?* https://www.cfr.org/backgrounder/what-canadas-immigration-policy#:~:text=In%201967%2C%20Ottawa%20introduced%20a,Latin%20America%20and%20the%20Caribbean

CIHI. (2017). *Infographic: Canada's senior population outlook: Uncharted territory.* A publication of the Canadian Institute for Health Information (CIHI). Available at https://www.cihi.ca/en/infographic-

---

[1] It was particularly difficult to find specific human capital initiatives in Canada online. This does not mean that there are no specific human capital initiatives. However, in comparison to the abundance sources on immigration in Canada, it could connote the relative importance placed on both sources of human capital.

[2] Herein lies a major potential advantage for African countries with large youthful populations.

canadas-seniors-population-outlook-uncharted-territory#:~:text=Over%20the%2020%20%20 years,sits%20at%20about%206.2%20million

Dhalla, I. A., & Tepper, J. (2018). Improving the quality of health care in Canada. *Canadian Medical Association Journal, 190*(39), E1162–E1167. https://doi.org/10.1503/cmaj.171045

Esho, E., & Verhoef, G. (2020). A holistic model of human capital for value creation and superior firm performance: The strategic factor market model. *Cogent Business & Management, 7*(1). https://doi.org/10.1080/23311975.2020.1728998

Fitzgerald, D. S., & Cook-Martín, D. (2014). *Culling the masses: The democratic origins of racist immigration policy in the Americas.* Harvard University Press.

Fraser Institute. (2022). *Canada's aging population – What does it mean for government finances?* Available at https://www.fraserinstitute.org/blogs/canadas-aging-population-what-does-it-mean-for-government-finances

Galor, O., & Weil, D. N. (2000). Population, technology, and growth: From malthusian stagnation to the demographic transition and beyond. *American Economic Review, 90*(4), 806–828.

Garrod, M., Vafaei, A., & Martin, L. (2020). The link between difficulty in accessing health care and health status in a Canadian context. *Health Services Insights, 2020*(13). https://doi.org/10.1177/1178632920977904

Goldin, C. (2001). The human capital century and American leadership: Virtues of the past. *Journal of Economic History, 61*, 263–291.

Goldin, C. (2016). Human capital. In C. Diebolt & M. Haupert (Eds.), *Handbook of cliometrics.* Springer.

Government of Canada. (2018). *The arrival of the Europeans: 17th century wars.* Available at https://www.canada.ca/en/department-national-defence/services/military-history/history-heritage/popular-books/aboriginal-people-canadian-military/arrival-europeans-17th-century-wars.html

Government of Canada. (2019). *Canada's health care system.* Available at https://www.canada.ca/en/health-canada/services/health-care-system/reports-publications/health-care-system/canada.html

Government of Canada. (2022a). *Notice – Supplementary information for the 2023–2025 immigration levels plan.* Available at https://www.canada.ca/en/immigration-refugees-citizenship/news/notices/supplementary-immigration-levels-2023-2025.html

Government of Canada. (2022b). *CIMM – Economic immigration – March 3, 2022.* Available at https://www.canada.ca/en/immigration-refugees-citizenship/corporate/transparency/committees/cimm-mar-03-2022/economic-immigration.html

Government of Canada. (2023a). *How express entry works.* Available at https://www.canada.ca/en/immigration-refugees-citizenship/services/immigrate-canada/express-entry/works.html

Government of Canada. (2023b). *Canada's health care system.* Available at https://www.canada.ca/en/health-canada/services/canada-health-care-system.html

Government of Canada. (2023c). *Notice – Supplementary information for the 2024–2026 immigration levels plan.* Available at https://www.canada.ca/en/immigration-refugees-citizenship/news/notices/supplementary-immigration-levels-2024-2026.html

Gu, W. (2023). *Accumulation of human capital in Canada, 1970 to 2020: An analysis of gender and the role of immigration* (Analytical Studies Branch Research Paper Series). Statistics Canada. Available at https://www150.statcan.gc.ca/n1/en/pub/11f0019m/11f0019m2023002-eng.pdf?st=JPpz-bFc

Hawthorne, L. (2013). Skilled enough? Employment outcomes for recent economic immigrants in Canada compared to Australia. In T. Triadafilopoulos (Ed.), *Wanted and welcome? Policies for highly skilled immigrants in comparative perspective* (pp. 219–256). Springer.

History Museum. (2023). *Historical overview of immigration to Canada.* Canadian Museum of History. Available at https://www.historymuseum.ca/cmc/exhibitions/tresors/immigration/imf0302e.html

Insead. (2022). *The global talent competitiveness index: The tectonics of talent: Is the world drifting towards increased talent inequalities?* Fontainebleau, France.

Kaushik, V., & Drolet, J. (2018). Settlement and integration needs of skilled workers in Canada. *Social Sciences, 7*(5), 76. https://doi.org/10.3390/socsci7050076

Kim, J., & Mahoney, J. T. (2007). Appropriating economic rents from resources: An integrative property rights and resource-based approach. *International Journal of Learning and Intellectual Capital, 4*(1/2), 11–28. https://doi.org/10.1504/IJLIC.2007.013820

Lake, M., & Reynolds, H. (2008). *Drawing the global colour line: White men's countries and the international challenge of racial equality.* Cambridge University Press.

Lim, S. S., Updike, R. L., Kaldjian, A. S., Barber, R. M., Cawling, K., York, H., Friedman, J., et al. (2018). Measuring human capital: A systematic analysis of 195 countries and territories, 1990–2016. *Lancet, 392*, 1217–1234.

Mankiw, G., Romer, D., & Weil, D. N. (1992). A contribution to the empirics of economic growth. *Quarterly Journal of Economics, 107*, 407–437.

Martin, D., Miller, A. P., Quesnel-Vallée, A., Caron, N. C., Vissandjée, B., & Marchildon, G. P. (2018). Canada's health-care system: Achieving its potential. *Lancet, 391*(10131), 1718–1735. https://doi.org/10.1016/S0140-6736(18)30181-8

Murphy, J. (2010). *The Settlement and integration needs of immigrants: A literature review.* A report of the Ottawa Local Immigration Partnership (OLIP). Available at https://olip-plio.ca/knowledge-base/wp-content/uploads/2013/03/Olip-Review-of-Literature-Final-EN.pdf

OECD. (2019). *Canada has the most comprehensive and elaborate migration system, but some challenges remain.* Available at https://web-archive.

oecd.org/2019-08-13/527484-canada-has-the-most-comprehensive-and-elaborate-migration-system-but-some-challenges-remain.htm

Paquet, M. (2014). The federalization of immigration and integration in Canada. *Canadian Journal of Political Science, 47*(3), 519–548.

Patrick, K., & Laupacis, A. (2023). A focus on access to health care in Canada. *Canadian Medical Association Journal, 195*(3), E123–E124. https://doi.org/10.1503/cmaj.230040

Reitz, J. G. (2012). The distinctiveness of Canadian immigration experience. *Patterns of Prejudice, 46*(5), 518–538. https://doi.org/10.1080/0031322X.2012.718168

Sharipov, I. (2015). *Contemporary economic growth models and theories: A literature review* (CES Working Papers, ISSN 2067-7693) (Vol. 7(3), pp. 759–773). Alexandru Ioan Cuza University of Iasi, Centre for European Studies.

Simmons, A. B. (2010). *Immigration and Canada: Global and transnational perspectives.* Canadian Scholar's Press.

Statistics Canada. (2016). *150 years of immigration in Canada.* Available at https://www150.statcan.gc.ca/n1/pub/11-630-x/11-630-x2016006-eng.htm

Statistics Canada. (2018). *Population growth: Migration increase overtakes natural increase.* Available at https://www150.statcan.gc.ca/n1/pub/11-630-x/11-630-x2014001-eng.htm

Statistics Canada. (2022a). *Immigrants make up the largest share of population in over 150 years and continue to shape who we are as Canadians.* Available at https://www150.statcan.gc.ca/n1/daily-quotidien/221026/dq221026a-eng.htm

Statistics Canada. (2022b). *Focus on geography series, 2021 census of population, 16-12-2022.* Available at https://www12.statcan.gc.ca/census-recensement/2021/as-sa/fogs-spg/page.cfm?topic=9&lang=E&dguid=2021A000011124

Statistics Canada. (2023). *Canada's population estimates: Record high population growth in 2022.* Available at https://www150.statcan.gc.ca/n1/daily-quotidien/230322/dq230322f-eng.htm

Thompson, J. H., & Weinfeld, M. (1995). Entry and exit: Canadian immigration policy in context. *The Annals of the American Academy of Political and Social Science, 538*(1), 185–198. https://doi.org/10.1177/0002716295538000015

Triadafilopoulos, T. (2021). The foundations, limits, and consequences of immigration exceptionalism in Canada. *American Review of Canadian Studies, 51*(1), 3–17. https://doi.org/10.1080/02722011.2021.1923150

Triadafilopoulos, T. (2022). Good and lucky: Explaining Canada's successful immigration policies. In E. Lindquist, M. Howlett, G. Skogstad, G. Tellier, & P. t'Hart (Eds.), *Policy success in Canada: Cases, lessons, challenges* (pp. 161–182). Oxford University Press.

UN. (2023). *Crude birth rate: Canada.* Available at https://population.un.org/dataportal/data/indicators/55/locations/124/start/1950/end/2023/table/pivotbylocation

University of Ottawa. (n.d.). *Linguistic history of Canada: Arrival of the Europeans and introduction of English and French.* Available at https://www.uottawa.ca/clmc/linguistic-history/arrival-europeans#:~:text=The%20first%20Europeans%20to%20come,not%20far%20from%20Saint%20Anthony

Van Dyk, L. (2023). *Canadian immigration acts and legislation: What do Canadian immigration rules tell us about Canada? Canadian Museum of Immigration at Pier 21.* Available at https://pier21.ca/research/immigration-history/canadian-immigration-acts-and-legislation?page=0

World Bank. (2022a). *GDP Per Capita (Current US$).* Available at https://genderdata.worldbank.org/indicators/ny-gdp-pcap-cd/

World Bank. (2022b). *Population 2022.* Available at https://databankfiles.worldbank.org/public/ddpext_download/POP.pdf

World Bank. (2023). *Life expectancy at birth, total (years).* Available at https://data.worldbank.org/indicator/SP.DYN.LE00.IN?view=chart

Part III

Human Capital in African Countries

State of Africa's Human Capital and Introduction to Part 3

9

Abstract

To present an overview of the state of human capital in African countries, some data and statistics on Africa's population and the performance of African countries on different available global human capital indexes were summarily analyzed in this chapter. On the average, with few exceptions of countries such as Seychelles and Mauritius, African countries generally rank poorly on many of the indexes. Despite the diverse methodological approaches and human capital measures and indicators used, most African countries came in at the bottom of the global rankings. Apart from the ranking positions, African countries also earned low scores on the indexes. Overall, these rankings reflect the poor state of human capital, its development, and acquisition, across African countries, and call for some urgent actions for the potential demographic dividends to be realized. This chapter concludes by introducing the final part of the book.

Keywords

Youthful population · Demographic dividend · Global indexes · Life expectancy · Human capital

Introduction

Africa is the second largest and second most populous continent in the world. Many of the 54 countries on the continent still face several developmental challenges and many are still classified as low-income countries. Of the total population of over 1.4 billion, an estimated 464 million people still live in extreme poverty in Africa (World Bank, 2024; World Bank PIP, 2024). The continent, especially Sub-Saharan Africa (SSA), has the youngest population in the world with about 70% of people in SSA currently under the age of 30 (Kamer, 2022; UN, 2022). The ten most populous countries on the continent are Nigeria, Ethiopia, Egypt, Democratic Republic of Congo, Tanzania, South Africa, Kenya, Uganda, Sudan, and Algeria. Each of these ten countries all have populations above 40 million (see Table 9.1). Despite the continent's huge population figures, only eight countries, Algeria, Angola, Egypt, Ethiopia, Kenya, Morocco, Nigeria, and South Africa, had gross domestic product (GDP) above US$100 billion in 2022, with Nigeria, Egypt, and South Africa as the top three countries.

A cursory analysis of the population, GDP and GDP per capita of the most populous countries show that the countries with the largest populations and GDP are not necessarily the countries with the highest GDP per capita. Seven of the countries with the highest GDP per capita do not have the highest GDP. Neither are these seven

Table 9.1 Population, GDP, and GDP per capita of African countries at a glance (2022)

S/N	Country	Subregion	Population (millions)	GDP (US$ billions)	GDP per capita (US$ thousands)
1.	Algeria	North Africa	44.90	191.91	4.27
2.	Angola	Central Africa	35.59	106.71	3.00
3.	Benin	West Africa	13.35	17.40	1.3
4.	Botswana	Southern Africa	2.63	20.35	7.74
5.	Burkina Faso	West Africa	22.67	18.88	0.83
6.	Burundi	East Africa	12.89	3.07	0.24
7.	Cape Verde	West Africa	0.59	2.30	3.90
8.	Cameroun	Central Africa	27.91	44.34	1.59
9	Central African	Central Africa	5.58	2.40	0.43
10.	Republic	Central Africa	17.72	12.70	0.72
11.	Chad	East Africa	0.84	1.24	1.48
12.	Comoros	Central Africa	99.01	58.07	0.59
13.	Congo (D.R.)	Central Africa	5.97	14.62	2.45
14.	Congo (Republic)	West Africa	28.16	70.02	2.49
15.	Cote d'Ivoire	East Africa	1.12	3.52	3.14
16.	Djibouti	North Africa	110.99	476.75	4.30
17.	Egypt	Central Africa	1.67	11.81	7.05
18.	Equatorial Guinea	East Africa	3.68	n/a	n/a
19.	Eritrea	Southern Africa	1.20	4.85	4.04
20.	Eswatini	East Africa	123.38	126.78	1.03
21.	Ethiopia	Central Africa	2.39	21.07	8.82
22.	Gabon	West Africa	2.71	2.27	0.84
23.	Gambia	West Africa	33.48	72.84	2.18
24.	Ghana	West Africa	13.86	21.23	1.53
25.	Guinea	West Africa	2.11	1.63	0.78
26.	Guinea-Bissau	East Africa	54.03	113.42	2.10
27.	Kenya	Southern Africa	2.31	2.55	1.11
28.	Lesotho	West Africa	5.30	4.00	0.75
29.	Liberia	North Africa	6.81	45.75	6.72
30.	Libya	East Africa	29.61	14.95	0.51
31.	Madagascar	East Africa	20.41	13.16	0.65
32.	Malawi	West Africa	22.59	18.83	0.83
33.	Mali	West Africa	4.74	10.38	2.19
34.	Mauritania	East Africa	1.26	12.90	10.22
35.	Mauritius	North Africa	37.46	134.18	3.53
36.	Morocco	East Africa	32.97	17.85	0.54
37.	Mozambique	Southern Africa	2.57	12.61	4.91
38.	Namibia	West Africa	26.21	13.97	0.53
39.	Niger	West Africa	218.54	477.39	2.18
40.	Nigeria	East Africa	13.78	13.31	0.97
41.	Rwanda	Central Africa	0.23	0.55	2.40
42.	Sao Tome and	West Africa	17.32	27.68	1.60
43.	Principe	East Africa	0.10	1.59	15.87
44.	Senegal	West Africa	8.61	3.97	0.46
45.	Seychelles	East Africa	17.60	8.13	0.46
46.	Sierra Leone	Southern Africa	59.89	405.87	6.78
47.	Somalia	East Africa	10.91	n/a	n/a
48.	South Africa	North Africa	46.87	51.66	1.10
49.	South Sudan	East Africa	65.50	75.71	1.19
50.	Sudan	West Africa	8.85	8.13	0.92
51.	Tanzania	North Africa	12.36	46.66	3.78
52.	Togo	East Africa	47.25	45.56	0.96
53.	Tunisia	Southern Africa	20.02	29.78	1.49
54.	Uganda	Southern Africa	16.32	20.68	1.27
	Zambia				
	Zimbabwe				

Source: World Development Indicators (2022a, b, c)

countries among the most populous countries on the continent. These seven countries, Seychelles, Mauritius, Gabon, Botswana, Equatorial Guinea, Libya, and Namibia, relative to their population size, are the most productive countries on the continent (see highlighted section of Table 9.2). The three countries that complete the top ten countries, Algeria, Egypt, and South Africa, are also among the ten most populous countries as well as the top ten countries with the highest GDP in Africa.

Data in Table 9.2 suggests that the economic productive capacity of most of the people in the most populous countries in Africa is low. Without needing to look at the various global human capital indexes, the data also suggests that the stock of human capital in the most populous countries is low either in quantity, quality, or both. Interestingly, the top countries with the highest GDP per capita such as Seychelles, Mauritius, Gabon, and Botswana (see Table 9.2) are also generally the highest ranked African countries on the different global human capital indexes—this is not mere coincidence.

Overview of Performance of African Countries Across Global Human Capital Indexes

Countries in SSA are usually grouped together and separated from North Africa on many of the global human capital indexes. However, like the previous chapters in Part 2, and because of the

focus of this book, a grouped analysis of the geographical context of Africa is preferred. Therefore, SSA and North Africa are presented together in this summary analysis.

African countries perform poorly across all global human capital indexes. On the first World Bank's Human Capital Index (WB HCI), published in 2018, 45 African countries were ranked among 157 countries. Only six African countries, Seychelles, Mauritius, Algeria, Kenya, Tunisia, and Morocco, were ranked in the top 100. Egypt and Gabon were ranked in the 104th and 110th positions, respectively. The other African countries on the index were ranked among the bottom 47 countries. Effectively, most of the lower rung of the index were occupied by African countries. The continent did not perform much better in the second edition of WB HCI of 2020. More countries were included in this edition, increasing the number to 174 countries from the previous 157 countries. Only 4 of the 46 African countries on the index were ranked in the top 100: Seychelles (42nd), Mauritius (58th), Kenya (93rd), and Algeria (98th). Again, similar to the previous index of 2018, most African countries made up the rear of the index as 38 of the bottom 50 countries were from Africa. Between 2013 and 2017, the World Economic Forum (WEF) published its annual Global Human Capital Index (GHCI), ranking different countries according to some measures of human capital. On all the years the index was published, only few African countries were ranked among the top 100 countries. Most African countries were also ranked at the rear.

A look at more recent human capital indexes reveals a similar situation. On the United Nations' Human Development Index (HDI) of 2021 and 2022, only Mauritius, Seychelles, Algeria, Egypt, and Tunisia were ranked among the top 100 countries. Of the 191 countries on the index, 39 of the lowest ranked 50 countries were African countries. Out of the 53[1] African countries on the index, 28 African countries were categorized as countries with "low human development," and only Mauritius was categorized under countries with "very high human development." Seychelles,

Table 9.2 Top ten African countries with the most population, GDP, and GDP per capita

S/N	Population	GDP	GDP per capita
1.	Nigeria	Nigeria	Seychelles
2.	Ethiopia	Egypt	Mauritius
3.	Egypt	South Africa	Gabon
4.	Congo	Algeria	Botswana
5.	(D.R.)	Morocco	Equatorial
6.	Tanzania	Ethiopia	Guinea
7.	South Africa	Kenya	South Africa
8.	Kenya	Angola	Libya
9.	Uganda	Tanzania	Namibia
10.	Sudan	Ghana	Egypt
	Algeria		Algeria

Source: World Development Indicators (2022a, b, c)

---

[1] Somalia is the only African country not on this index.

Algeria, Egypt, Tunisia, Libya, South Africa, and Gabon were the only African countries categorized as countries with "high human development." The ranking of African countries on the most current edition of Global Talent Competitiveness Index (GTCI), published by Insead Business School, France, in collaboration with Portulans Institute, USA, and Human Capital Leadership Institute, Singapore, is not much different. In the latest 2023 edition of the index, Mauritius, the topmost African country, was ranked in the 51st position, while South Africa came in second at the 68th position. Ten other African countries, Botswana (73rd), Cape Verde (78th), Egypt (88th), Tunisia (92nd), Namibia (93rd), Ghana (95th), Gambia (97th), Kenya (98th), Morocco (99th), and Eswatini (100th), were ranked among the top 100 countries globally.

The main concern is not necessarily the general low rankings of African countries on the global indexes for human capital. Rather, it is the low scores of African countries on the diverse measures of human capital used across the indexes. African countries have similar performances on more "academically" inclined human capital indexes which are not necessarily focused on the rankings of countries. On the period measure of human capital computed by Lim et al. (2018), for example, Libya, Tunisia, Seychelles, Algeria, Mauritius, Egypt, Gabon, and Morocco had measures of 10 years and above. All other 46 countries on the continent had period human capital of less than 10 years.[2] This is one human capital index that ranked all 54 African countries. For comparison, the top countries globally on this index had period human capital scores that ranged from 25 to 28 years.

The utility of these global human capital indexes is not necessarily in their ranking positions and placements of countries. It is possible to argue that the low placing of most of the ranked

African countries on these indexes does not really matter. This argument would be viable if all the countries on the indexes had good or great human capital scores. It is important to reiterate that the poor ranking of African countries is worrisome not just because of their low placements on these indexes but majorly because of their low scores in the diverse measures of human capital used on the different indexes. Rankings actually matter less if scores in themselves reflect good performance. However, it is also important to note the limitations of these global indexes. Although all the indexes generally measure human capital in one form or the other, the purpose, data, and methodology of each index is different. Consequently, the indexes serve majorly as a guide, and their optimal utility lies in using the indexes jointly rather than a focus on only one index.

Other Indicators of Human Capital of African Countries

The general poor performance of African countries across the human capital indexes is not accidental. A delve down into some of the indicators that constitute the basic building blocks of most of the indexes reveals a clearer picture of the state of Africa's human capital. The number of children that are out of school in Africa, especially in SSA, is still staggering. While the total number of out-of-school children and youths between the ages of 6 and 18 years worldwide fell from 400 million in 2000 to 250 million in 2022, that of SSA has risen to almost 100 million (Dharamshi et al., 2022; UNESCO, 2023). Twenty percent of children are out of primary school, and about 60% of youths between 15 and 17 years old in SSA are out of school (Klapper & Panchamia, 2023). The number of compulsory schooling years across African countries is still low. From available data, as at 2015, only 39 African countries had adopted 7 years of compulsory primary school education (Carr, 2022).

These figures and statistics are pointers to the state of education systems in African countries. Formal education is still exclusive. Many chil-

---

[2]Lim and colleagues refer to their measure of period human capital as the expected number of years lived between 20 and 64 years that takes into account educational attainment, learning or education quality, and functional health status of countries.

dren and youths either do not have or cannot access formal education. Without formal primary, secondary, and tertiary education, the basic foundations of human capital development are surely weak. Indigenous knowledge systems are also rather weak, largely informal, fragmented, and unregulated. At an era when global competition requires much more than having a workforce with tertiary education (Porter, 1990), compulsory primary and secondary school education has still not been achieved by many African countries. Consequently, the stock of human capital, regardless of its quality, is still rather low. The basic literacy required to strive in current modern-day societies with digital technologies has gone beyond basic literacy, the abilities to read and write, skills taught in formal primary school education, to digital literacy, the ability to access and communicate using digital technologies. Africa's poor human capital development is further compounded by the weak health-care systems of many African countries. Many countries' health-care systems still operate out-of-pocket payment systems; with the exception of few countries such as Rwanda, universal health-care systems are still the exception rather than the norm in Africa (Azevedo, 2017; Ly et al., 2022; UNDPI, 2017). With poor education and health-care systems, the poor rankings of many African countries on global human capital indexes should come as no surprise.

The ages beginning from middle adolescence of about 15–65 years are generally recognized as the most productive age of a person. This also translates to the context of countries. The most productive people in a country's population are those between this age range. The percentage of the population between this age range in Africa varies from 70.9% in Mauritius to 48.75% in Niger (see Table 9.3). For all African countries except Mauritius, the percentage of people below 65 years is over 90%, going as high as over 97% in many countries. Currently, the working-age population, those between 15 and 64 years, range from 48.75% in Niger to 70.9% in Mauritius. These statistics, in addition to the large percentage of people below 15 years, coupled with cur-

rent high birth rates in most African countries reflect not only a youthful population but also a population that is set to continue to grow and remain youthful, not only in the near future but far into the future. The average birth rate for an African woman is 4.8 children, a rate double that of the world's average of 2.4 children per woman (World Bank, 2021). On the flip positive side, a young and growing youthful population is also a potential for African countries to reap demographic dividends.

However, demographic dividends, an increase in economic growth that follows a change in a country's population, more specifically, an increase in the working-age population (Bloom et al., 2003; Ssewamala, 2015; Zhou et al., 2023), can only accrue to countries with the requisite human capital. Huge working-age populations do not automatically translate to democratic dividends. The implicit assumption in dividends resulting from an increase in the working-age population of a country is that the people within the requisite age groups are gainfully employed. In other words, there are gains to having a huge population of working-age people who are utilizing and deploying their human capital in economically productive activities.

From the perspective of some, for example, Zhou et al. (2023), demographic dividends are perhaps more actually like education dividends. I align strongly with this perspective. Demographic dividends can actually be referred to as human capital dividends because economic growth from increases in working-age population results mainly from a working-age population that is properly equipped with the *necessary* human capital. In a global economy that is increasingly becoming digitalized because of the rapid advent of digital technologies, the necessary human capital should ordinarily be one that can be usefully and gainfully deployed in a digital economy. Consequently, despite the huge potential the large youthful population numbers in Africa portend for the yields from demographic dividends, such potential will remain only a potential without adequate investments in human capital development and accumulation.

Table 9.3 Percentage of population of different age ranges

Country	Population age 0–14 years (%)	Population age 15–64 years (%)	Population age 0–64 years (%)	Population ages 65 years and above (%)	Fertility rate (no. of births per woman) 2021
Algeria	30.63	62.97	**93.61**	6.39	2.889
Angola	45.02	52.38	**97.40**	2.60	5.304
Benin	42.43	54.50	**96.94**	3.06	4.973
Botswana	32.60	63.74	**96.35**	3.65	2.791
Burkina Faso	43.72	53.74	**97.47**	2.53	4.772
Burundi	45.77	51.75	**97.52**	2.48	5.078
Cape Verde	26.18	68.27	**94.45**	5.55	1.896
Cameroun	42.21	55.12	**97.33**	2.67	4.463
Central African	48.13	49.35	**97.49**	2.51	5.978
Republic	47.51	50.48	**97.99**	2.01	6.255
Chad	38.04	57.69	**95.72**	4.28	3.978
Comoros	46.54	50.54	**97.08**	2.92	6.156
Congo (D.R.)	41.05	56.23	**97.28**	2.72	4.171
Congo (Republic)	41.48	56.13	**97.60**	2.40	4.418
Cote d'Ivoire	30.39	65.07	**95.46**	4.54	2.804
Djibouti	32.86	62.31	**95.17**	4.83	2.917
Egypt	38.49	58.38	**96.88**	3.12	4.266
Equatorial Guinea	39.20	56.79	**95.99**	4.01	3.867
Eritrea	34.69	61.30	**96.00**	4.00	2.839
Eswatini	39.63	57.23	**96.86**	3.14	4.159
Ethiopia	36.28	59.83	**96.11**	3.89	3.491
Gabon	43.06	54.52	**97.57**	2.43	4.684
Gambia	36.94	59.51	**96.45**	3.55	3.563
Ghana	41.54	55.13	**96.68**	3.32	4.399
Guinea	40.11	57.07	**97.18**	2.82	4.005
Guinea-Bissau	37.81	59.32	**97.13**	2.87	3.335
Kenya	33.98	61.82	**95.80**	4.20	3.018
Lesotho	40.52	56.17	**96.69**	3.31	4.089
Liberia	28.29	66.84	**95.14**	4.86	2.462
Libya	39.07	57.58	**96.65**	3.35	3.851
Madagascar	42.59	54.80	**97.39**	2.61	3.917
Malawi	47.19	50.43	**97.62**	2.38	5.956
Mali	41.61	55.17	**96.78**	3.22	4.398
Mauritania	16.31	70.90	**87.21**	12.79	1.41
Mauritius	26.58	65.70	**92.28**	7.72	2.328
Morocco	43.52	53.91	**97.43**	2.57	4.644
Mozambique	36.24	59.79	**96.03**	3.97	3.303
Namibia	48.85	48.75	**97.60**	2.40	6.82
Niger	43.04	53.98	**97.03**	2.97	5.237
Nigeria	38.47	58.34	**96.80**	3.20	3.821
Rwanda	39.47	56.77	**96.24**	3.76	3.823
Sao Tome and	41.45	55.41	**96.86**	3.14	4.387
Principe	23.10	68.75	**91.85**	8.15	2.46
Senegal	38.98	57.88	**96.86**	3.14	3.978
Seychelles	47.16	50.27	**97.43**	2.57	6.312
Sierra Leone	28.55	65.56	**94.11**	5.89	2.374
Somalia	43.87	53.24	**97.11**	2.89	4.469
South Africa	40.93	55.57	**96.50**	3.50	4.457
South Sudan	43.36	53.54	**96.90**	3.10	4.726
Sudan	39.96	56.90	**96.87**	3.13	4.257
Tanzania	24.84	66.14	**90.98**	9.02	2.086
Togo	44.78	53.52	**98.31**	1.69	4.585
Tunisia	42.86	55.40	**98.25**	1.75	4.308
Uganda	40.63	56.04	**96.68**	3.32	3.491
Zambia					
Zimbabwe					

Source: World Development Indicators

It is important to note that this focus on formal education that is traditionally through classrooms in a country's educational system is only a simplistic one. Human capital is a multidimensional component that consists of knowledge from education and other sources and that health is also one of its fundamental components. However, health is usually implicitly subsumed in the analytical focus on formal education as people have to of necessity be healthy to develop or deploy their human capital.

Reliability of the Human Capital Indexes

The human capital rankings can be said to be a correct reflection of the state of human capital in African countries. Liu and Fraumeni (2020) show a high degree of correlation between different human capital indexes. The correlation between World Bank's HCI, United Nations HDI, and World Economic Forum (WEF)'s GHCI ranged from 0.85 to 0.94 (Liu & Fraumeni, 2020). Despite the different methodological approaches adopted by the different human capital indexes, the scoring for countries, and order of rankings especially, are similar to a large extent across the indexes. Remarkably, despite the diversity of the indicators used as human capital measures, the lowest ranked African countries are somewhat consistent, to a large degree, across all the human capital indexes. So are the top-ranked African countries. In fact, generally, Seychelles and Mauritius are the two best-ranked African countries across many of the indexes. These two countries also not only consistently scored above average human capital scores; they also ranked above some developed countries on some indexes. Clearly, these two countries seem to be doing somethings about their human capital that is worth taking note of. It is no coincidence that Mauritius and Seychelles are also the countries with the highest GDP per capita in Africa by a huge margin. While Seychelles and Mauritius had GDP per capita of over US$15,000 and

US$10,000, respectively, Gabon and Botswana, the next two countries, had GDP per capita of US$8000 and US$7000, respectively. Although several factors can be masked within GDP per capita, in terms of human capital and its development, GDP per capita is a general reflection of the productivity level of countries, given their population. Interestingly, Gabon and Botswana also generally rank among the top African countries on different human capital indexes.

The relatively well-placed rankings of Seychelles and Mauritius seem to be the results of deliberate strategic human capital development plans. In 2018, Seychelles launched a new 5-year strategic policy on human capital development with support from the African Development Bank (ADB, 2019). This is in addition to having an agency dedicated to human capital development in the country, the Agency for National Human Resources Development (ANHRD), and formal laid out multi-year plans for national development. Similarly, Mauritius has a national agency, the Human Resource Development Council (HRDC), responsible for human capital development. HRDC develops national human capital development strategies in its development plan, the National Human Resource Development Plan (NHRDP), which was first launched in 2007 (HRDC, 2020). The NHRDPs aim to provide the policy frameworks for education, trainings, and career progression. Human capital was also explicitly recognized as an important element in the country's quest for development in its first national development plan of 1971 (Bunwaree, 2001). Clearly, the top rankings of these two countries among African countries are not coincidental. However, human capital indexes for Seychelles and Mauritius show that their human capital are still way behind those of some countries in other world regions and still below targeted levels. Summarily, from the various human capital indexes, data, and statistics, Africa's human capital levels pale in comparison to that of other continents and to achievable benchmarks given the continent's huge youthful and working-age population.

Other Indices of Human Capital of African Countries: Some Comparative Analyses

A look at other indicators of human capital of African countries reveals trends similar to that revealed by the global indexes. Overall, Africa, as a continent, slacks visibly behind other world regions in the average years of schooling of the working-age population (Fig. 9.1). The literacy rates of the working-age population for African countries also trail that of countries in other world regions (Fig. 9.2).

Another indicator of a country's human capital is the average life expectancy at birth. This is meant, especially, to reflect the health status of the average person in the population. Unsurprisingly, as health is a component of human capita, there is also evidence that it can be a reflection of the level of human capital in countries (Hansen, 2013; Hoque et al., 2019; Kotschy, 2021; Vu, 2023). From Figs. 9.3 and 9.4, the average life expectancy at birth of African countries trails behind that of other world regions by a far margin. The grouping of North African countries with countries in the Middle East in Fig. 9.3 masks the generally low average life expectancy in countries in North Africa. Summarily, the life expectancy in African countries is not only low across all subregions in

Africa; it is well below that of other world regions—this is more visible in Fig. 9.4.

The above graphical analyses give further credence to the global human capital indexes. One may argue that the indicators make up some of those included in the global indexes. Consequently, a correlation between the two is not surprising. However, only some of the global indexes contain the above indicators in some forms.

Government Expenditure on Education In 2015, the United Nations Educational Scientific and Cultural Organizations (UNESCO) recommended that member countries should allocate 4–6% of their GDP or 15–20% of their budgets to education (UNESCO, 2021). A general survey of governments' expenditure on education as a percentage of GDP shows a wide variation across African countries and indeed across countries globally. While the average for African countries has hovered around the recommended minimum threshold of 4%, countries such as Botswana, Lesotho, Namibia, and Sierra Leone have spent as much as 7–9% of their GDP on education. Some other countries have spent as little as 1% of their GDP on education. However, government spending on education in relation to GDP in African countries is generally within that of other countries globally (see Fig. 9.5).

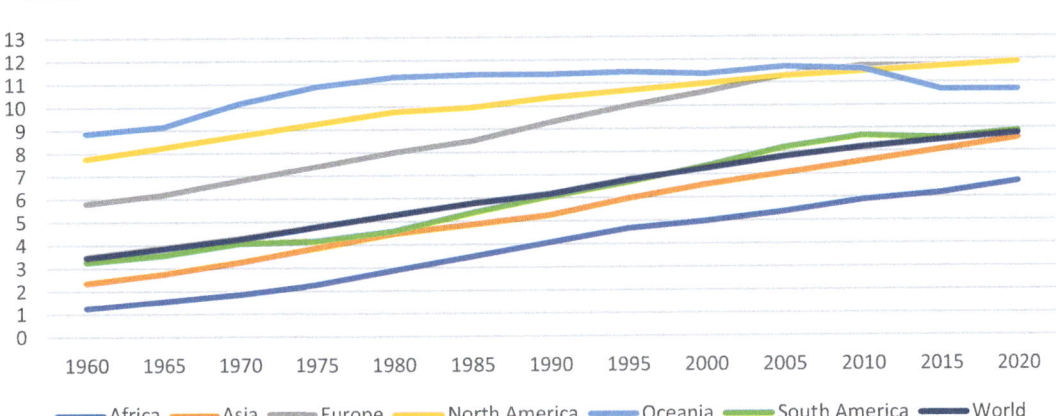

Fig. 9.1 Average years of schooling: 15–64 years. (Source of data for graph: https://ourworldindata.org/, Barro and Lee (2015), and Lee and Lee (2016))

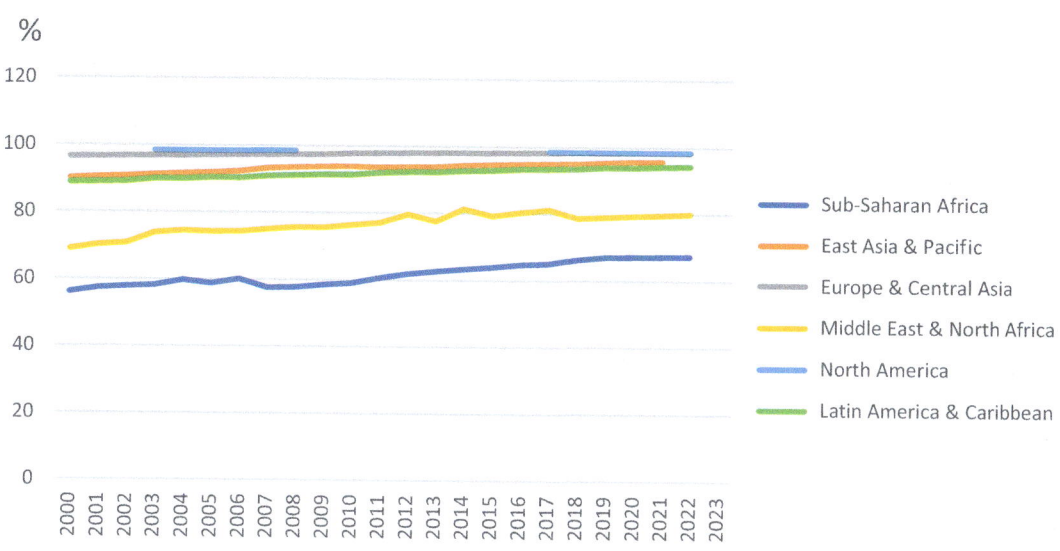

Fig. 9.2 Literacy rates of population 15 years and above. (Source of data for graph: World Development Indicators (2022d))

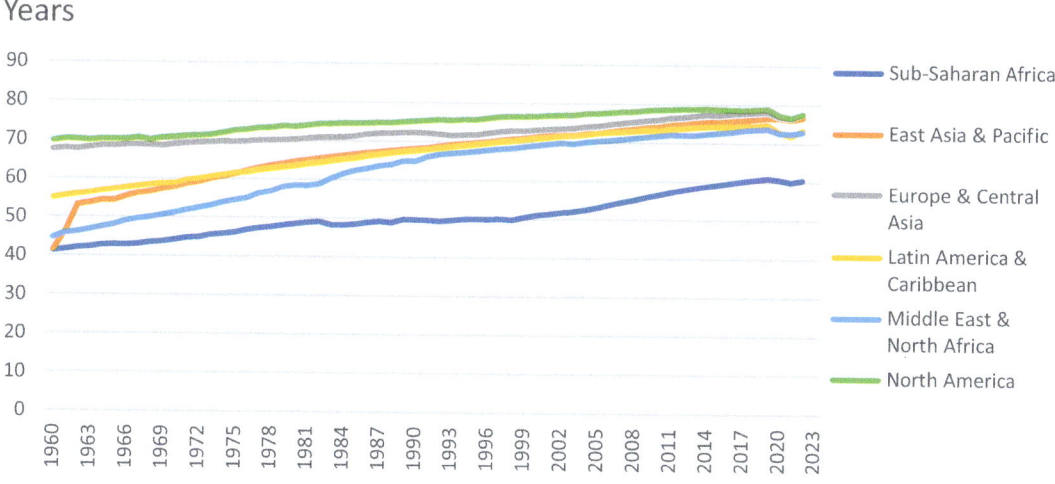

Fig. 9.3 Average life expectancy at birth. (Source of data for graph: World Development Indicators (2022e))

Most African countries have actually met at least one of UNESCO's recommendations (AfDB, 2020). The challenge is not with the amounts spent which, given the GDP of African countries, is arguably small compared to that of other more advanced countries. The challenge is with the spending efficiency of government expenditure on education which is generally quite low for many countries on the continent (Esho & Verhoef, 2022; Miningou, 2019; Sikayena et al., 2022).

Africa's Human Capital Development: Historical Context and Future Pathways

A discussion on the current state of human capital in Africa would be incomplete without any reference to the historical trajectory of the continent's human capital development. Despite the current state of the continent's stock of human capital in comparison with the inherent potentials given the data on population figures, history

Years

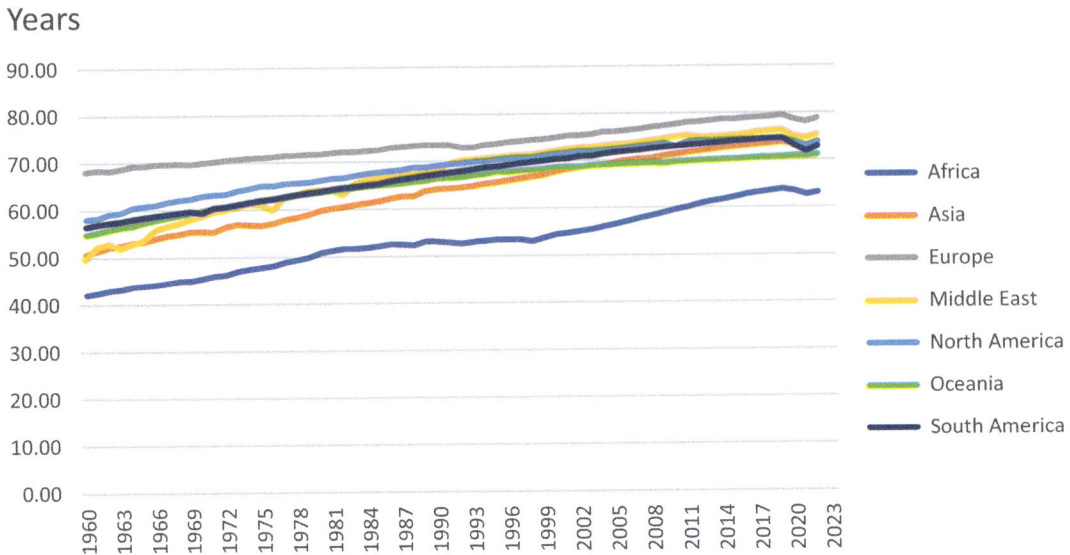

Fig. 9.4 Average life expectancy at birth (author's revised continental grouping). (Source of data for graph: World Development Indicators (2022e))

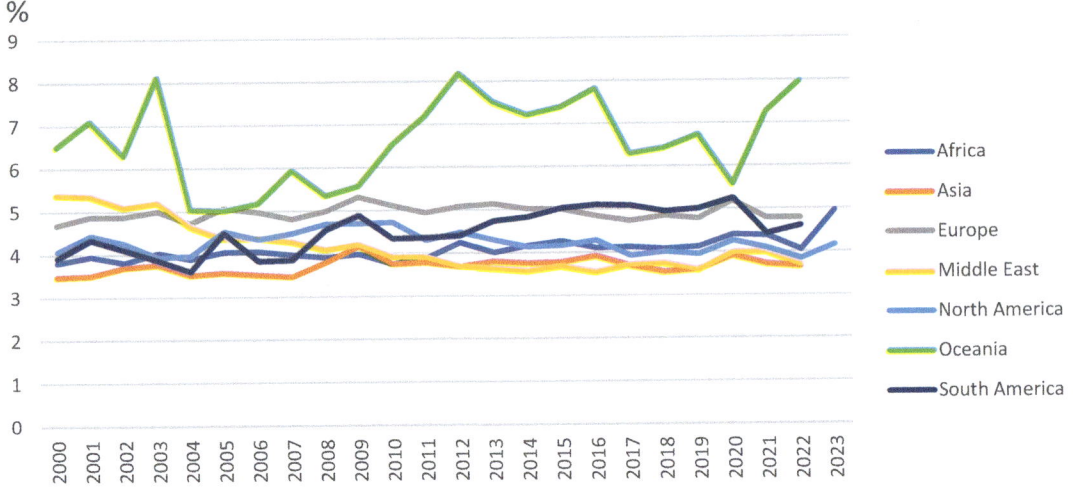

Fig. 9.5 Total government expenditure on education as a percentage of GDP. (Source of data for graph: World Development Indicators (2022f))

helps to put these in perspective. Along with this perspective are also glimmers of hope that the current general dire state of human capital in African countries has bright prospects for change.

Precolonial Africa had systems of indigenous knowledge that served as forms of educational systems (Emeagwali, 2014; Heto & Mino, 2023; Kaya & Seleti, 2013). Apprenticeships that served as a means of training in vocations and occupations were common in many African societies (Adeola, 2021; Oyewunmi et al., 2021), and the organization of business firms were majorly within families and clans (Verhoef, 2017). These point to forms of organizations and organizing that facilitated human capital development in precolonial Africa, and the of stock human capital developed was quite apt for the social and economic needs of the times.

Colonialization brought along western form of formal education which has since become the norm of formal education systems worldwide and a basic source of human capital for contemporary economies. Different fundamental aspects of the gauge for assessing human capital development rest on western-style education systems which have been adapted in various ways globally by countries. School enrollment and completion rates, for examples, are based on data derived from the various adaptations of western formal education systems. Despite the current poor state of human capital in Africa, unbelievably, dramatic progress has been made in human capital development. Between 1971 and 2015, for example, the proportion of children completing primary school education in Sub-Saharan Africa actually rose from 27% in 1971 to 67% in 2015, and on the average, the percentage of people completing lower secondary school education rose from a mere 5% to 40% within the same period (Evans & Acosta, 2021; World Bank, 2020). These few simple statistics show how far African countries have come in developing human capital for contemporary economic times and the possibilities.

In a bid to promote human capital development in Africa, the World Bank, in 2019, launched the Africa Human Capital Plan (HCP). Through the HCP, world bank spent a total sum of US$8.96 billion on various human capital development projects and initiatives on the continent in 2021 (World Bank, 2021). Human capital development is complex, and some of the different complexities result from the fact that investments in people, investment in their human capital, can be direct and or indirect. For example, while investments in education and health care can be regarded as direct development of human capital, other investments into supporting infrastructure such as transportation, water systems, and digital infrastructure that enable people to access formal education and health care are not direct investments in human capital. However, these types of investments can be regarded as indirect investments in people and their human capital, all things being equal. Consequently, the pathways toward human capital development in Africa present enormous challenges that require huge financial and nonfinancial investments in direct and indirect investments in people and their human capital.

In the remaining chapters of this book, some practical recommendations on human capital development, based on a sound theoretical framework, are provided. In Chap. 10, the next chapter, the generic strategic approaches which African countries, and indeed all countries, can utilize for human capital accumulation are presented. Chapter 11 provides some specific in-depth recommendations and suggestions than can be adopted and adapted by African countries to increase their human capital. Chapter 12 concludes this third part of the book and the book.

Conclusion

Available data and statistics point to the current poor state of human capital in Africa. Despite the huge strides that have been made since the immediate past colonial era, most African countries still rank poorly on the global human capital indexes. Interestingly, alongside these dismal statistics is a huge and growing young and youthful population, one that has been estimated to be the youngest population on earth in the near future. The promise of demographic dividends that Africa's youthful population portends cannot be realized without appropriate investments in human capital development. While the World Bank's human capital project might indeed be laudable, the onus is on African governments and Africans themselves, to take up the challenge of developing the continent's most abundant and crucial resource—the people and their human capital.

References

ADB. (2019). *African Development Bank (ADB) Report. Seychelles launches new HR policy and strategy 2018–2022.* https://www.afdb.org/en/news-and-events/seychelles-launches-new-hr-policy-and-strategy-2018-2022-18916

Adeola, O. (2021). The Igbo traditional business school (I-TBS): An introduction. In O. Adeola (Ed.), *Indigenous Africa enterprise: The Igbo traditional business school* (Advanced series in management) (Vol. 26, pp. 3–12). Emerald Publishing.

AfDB. (2020). *African economic outlook 2020*. African Development Bank. Available at https://www.afdb.org/en/documents/african-economic-outlook-2020. Accessed 04.2020.

Azevedo, M. J. (2017). The state of health system(s) in Africa: Challenges and opportunities. In *Historical perspectives on the state of health and health systems in Africa, Volume II. African histories and modernities*. Palgrave Macmillan. https://doi.org/10.1007/978-3-319-32564-4_1

Barro & Lee (2015). Lee & Lee (2016). With major processing by Our World in Data. "Average years of schooling" [dataset]. Barro and Lee, "Projections of Educational Attainment"; Lee and Lee, "Human Capital in the Long Run" [original data]. Retrieved August 30, 2024 from https://ourworldindata.org/grapher/mean-years-of-schooling-long-run

Bloom, D. E., Canning, D., & Sevila, J. (2003). The debate over the effects of population growth on economic growth. In *The demographic dividend: A new perspective on the economic consequences of population change* (pp. 1–24). Rand Corporation.

Bunwaree, S. (2001). The marginal in the miracle: Human capital in Mauritius. *International Journal of Educational Development, 21*(3), 257–271. https://doi.org/10.1016/S0738-0593(00)00033-X

Carr, O. G. (2022). Promoting priorities: Explaining the adoption of compulsory schooling laws in Africa. *International Journal of Educational Development, 88*, 102523. https://doi.org/10.1016/j.ijedudev.2021.102523

Dharamshi, A., Barakat, B., Antoninis, M., Montoya, S., & UNESCO. (2022). *A Bayesian cohort model for estimating SDG indicator 4.1.4: Out-of School rates*. https://www.unesco.org/gem-report/sites/default/files/medias/fichiers/2022/08/OOS_Proposal.pdf

Emeagwali, G. (2014). Intersections between Africa's Indigenous knowledge systems and history. In G. Emeagwali & G. J. S. Dei (Eds.), *African Indigenous knowledge and the disciplines. Anti-colonial educational perspectives for transformative change*. Sense Publishers. https://doi.org/10.1007/978-94-6209-770-4_1

Esho, E., & Verhoef, G. (2022). Reaping the benefits of African Continental Free Trade Agreement (AfCFTA): The role of human capital development. *Africa Review, 15*(1), 1–23.

Evans, D. K., & Acosta, A. M. (2021). Education in Africa: What are we learning? *Journal of African Economies, 30*(1), 13–54. https://doi.org/10.1093/jae/ejaa009

Hansen, C. W. (2013). Life expectancy and human capital: Evidence from the international epidemiological transition. *Journal of Health Economics, 32*(6), 1142–1152. https://doi.org/10.1016/j.jhealeco.2013.09.011

Heto, P. P.-K., & Mino, T. (2023). (Dis)continuity of African Indigenous knowledge. *AlterNative: An International Journal of Indigenous Peoples, 19*(1), 71–79. https://doi.org/10.1177/11771801221138304

Hoque, M. M., King, E. M., Montenegro, C. E., & Orazem, P. F. (2019). Revisiting the relationship between longevity and lifetime education: Global evidence from 919 surveys. *Journal of Population Economics, 32*(2), 551–589. https://doi.org/10.1007/s00148-018-0717-9

HRDC. (2020). *National Human Resource Development Plan (NHRDP)*. Available at https://www.hrdc.mu/index.php/nhrdp

Kamer, L. (2022). *Median age in Africa 2022, by country*. Statista Report.

Kaya, H. O., & Seleti, Y. N. (2013). African indigenous knowledge systems and relevance of higher education in South Africa. *The International Education Journal: Comparative Perspectives, 12*(1), 30–44.

Klapper, L., & Panchamia, M. V. (2023). *The high price of education in Sub-Saharan Africa*. World Bank Blog. https://blogs.worldbank.org/developmenttalk/high-price-education-sub-saharan-africa

Kotschy, R. (2021). Health dynamics shape life-cycle incomes. *Journal of Health Economics, 75*, 102398. https://doi.org/10.1016/j.jhealeco.2020.102398

Lim, S. S., Updike, R. L., Kaldjian, A. S., Barber, R. M., Cawling, K., York, H., Friedman, J., et al. (2018). Measuring human capital: A systematic analysis of 195 countries and territories, 1990–2016. *Lancet, 392*, 1217–1234. https://doi.org/10.1016/S0140-6736(18)31941-X

Liu, G., & Fraumeni, B. M. (2020). *A brief introduction to human capital measures*. IZA Institute of Labor Economics, Discussion Paper Series, IZA DP No. 13494.

Ly, M. S., Bassoum, O., & Faye, A. (2022). Universal health insurance in Africa: A narrative review of the literature on institutional models. *BMJ Global Health, 22*(7), e008219. https://doi.org/10.1136/bmjgh-2021-008219

Miningou, E. W. (2019). *Quality education and the efficiency of public expenditure: A cross-country comparative analysis*. World Bank Policy Research Working Paper No. 9077. Available at SSRN: https://ssrn.com/abstract=3501936

Oyewunmi, A. E., Oyewunmi, O. A., & Moses, C. L. (2021). Igba-Boi: historical transitions of the Igbo apprenticeship model. In O. Adeola (Ed.), *Indigenous Africa enterprise: The Igbo traditional business school* (Advanced series in management) (Vol. 26, pp. 13–25). Emerald Publishing.

Porter, M. (1990). The competitive advantage of nations. *Harvard Business Review*, March–April, pp. 73–91. Available at https://economie.ens.psl.eu/IMG/pdf/porter_1990_-_the_competitive_advantage_of_nations.pdf

Sikayena, I., Bentum-Ennin, I., Andoh, F. K., & Asravor, R. (2022). Efficiency of public spending on human capital in Africa. *Cogent Economics & Finance, 10*(1). https://doi.org/10.1080/23322039.2022.2140905

Ssewamala, F. M. (2015). Optimizing the 'demographic dividend' in young developing countries: The role of contractual savings and insurance for financing education. *International Journal of Social Welfare, 24*(3), 248–262. https://doi.org/10.1111/ijsw.12131

UN. (2022). *Young people's potential, the key to Africa's sustainable development.* Available at https://www.un.org/ohrlls/news/young-people%E2%80%99s-potential-key-africa%E2%80%99s-sustainable-development

UNDPI. (2017). Health care systems: Time for a rethink. *Africa Renewal, 30*(3), 4–19.

UNESCO. (2021). *UNESCO member states unite to increase investment in education.* Available at https://www.unesco.org/en/articles/unesco-member-states-unite-increase-investment-education

UNESCO. (2023). *Out-of-school rate.* Available at https://education-estimates.org/out-of-school/

Verhoef, G. (2017). *The history of business in Africa: Complex discontinuity to emerging markets.* Springer International.

Vu, T. V. (2023). Life expectancy and human capital: New empirical evidence. *Health Economics, 32*(2), 395–412. https://doi.org/10.1002/hec.4626

World Bank. (2020). *World Development Indicators.* Available at https://databank.worldbank.org/source/world-development-indicators

World Bank. (2021). *Investing in people for a resilient and inclusive recovery.* Africa Human Capital Plan, Year Two Progress Report.

World Bank. (2024). *The World Bank in Africa: Overview.* Available at https://www.worldbank.org/en/region/afr/overview

World Bank PIP. (2024). *The World Bank Poverty and Inequality Platform (PIP).* Available at https://pip.worldbank.org/home

World Development Indicators. (2022a). *Population, total.* Available at https://data.worldbank.org/indicator/SP.POP.TOTL

World Development Indicators. (2022b). *GDP (Current US$).* World Bank National Accounts, and OECD National Accounts Data Files. Available at https://data.worldbank.org/indicator/NY.GDP.MKTP.CD

World Development Indicators. (2022c). *GDP per capita (current US$).* World Bank National Accounts, and OECD National Accounts Data Files. Available at https://data.worldbank.org/indicator/NY.GDP.PCAP.CD

World Development Indicators. (2022d). *Literacy rate, adult total (% of people ages 15 and above).* UNESCO Institute of Statistics (UIS), UIS Bulk Data Download Service. Available at https://data.worldbank.org/indicator/SE.ADT.LITR.ZS

World Development Indicators. (2022e). *Life expectancy at birth, total (years).* United Nations Population Division, World Population Prospects: 2022 Revision. Available at https://data.worldbank.org/indicator/SP.DYN.LE00.IN

World Development Indicators. (2022f). *Government expenditure on education, total (% of GDP).* UNESCO Institute for Statistics (UIS), UIS Bulk Data Download Service. Available at https://data.worldbank.org/indicator/SE.XPD.TOTL.GD.ZS

Zhou, J., Deng, J., Li, L., & Wang, S. (2023). The demographic dividend or the education dividend? Evidence from China's growth. *Sustainability, 15*(9), 7309. https://doi.org/10.3390/su15097309

Generic Strategic Approaches to Human Capital Development and Accumulation

10

Abstract

Using human capital theory, resource-based theory (RBT), and transaction cost economies, this chapter presents three generic approaches that can be adopted by African countries to develop and accumulate human capital. Countries can either accumulate human capital internally through human capital development such as their formal education system, workplace training and development, and developing their indigenous knowledge system and informal education; externally through strategic immigration, foreign direct investments (FDI), and foreign study; or through using a hybrid of the two major approaches. In reality, countries generally use a hybrid approach as it is the most realistic approach to human capital accumulation. However, the extent to which the hybrid approach leans on either of the two other approaches depends on the needs and situation of each country. The crucial factor is that human capital accumulation is not left to chance but is planned for in a strategic manner. The chapter presents an in-depth discussion of the different routes within each strategic approach.

Keywords

Brain drain · Diaspora remittance · Micro–macro paradox · Human capital · RBT

Introduction

The knowledge, skills, abilities, and other characteristics (KSAOs) of individuals that constitute human capital require investments. This chapter provides some answers to how human capital investments can be made both within the context of the discussion on developing Africa's human capital and generally within the wider context of human capital development and accumulation. Unlike many other investments, human capital investments usually take a while, a long time, to yield visible results. In fact, investments in human capital usually do not begin to yield returns until the KSAOs that have been accumulated over time are utilized by being put to use in work. While individuals can accumulate human capital by and for themselves, a country's stock of human capital is accumulated through individuals, usually primarily through the citizens and residents within the country. The productivity of a country's stock of human capital generally depends on factors such as the nature of the stock of human capital in terms of type, quantity, quality, and several other national and global contextual factors. While global contextual factors may be beyond the direct control of countries, national facilitating factors are usually within their control.

Having an adequate stock of human capital is a prerequisite for productivity and national global competitiveness. However, what can be regarded as an adequate stock of human capital is relative. The notion of an adequate stock of human capital is that a country has "enough" people with the requisite human capital—in quantity—and that these people have accumulated the right quality of human capital. Consequently, much like firms and other forms of organizations, countries need to have strategic approaches to human capital accumulation, because countries are basically a form of "organization." Countries need to build up their stock of human capital by having people with the capacities for economic production and for engendering global competitiveness. However, having the adequate stock of human capital in a country does not just happen. It is not accidental. A country's decision to accumulate human capital needs to be deliberate, planned, and strategic in order to be effective. It's not one that should be left to chance especially when its current stock of human capital is low, as is the definite case for most African countries.

Using the rudiments of human capital theory, resource-based theory, and transaction cost economics, this chapter develops three generic strategic approaches to countries' human capital accumulation. Although these theories are primarily directed at firms, some aspects of the theories are nonetheless relevant to gaining insights into countries' strategic approaches to human capital accumulation and development.

Theoretical Foundation

Human Capital Theory This book is on human capital, and much of its contents reflect the tenets of human capital theory (e.g., Becker, 1964, 1993; Schultz, 1961). More explicitly, the main proposition of human capital theory is that investments in people are a form of capital capable of yielding returns like any other form of capital. Investment in people can range from investments in formal and informal education, health care, migration, apprenticeships, all forms of learning, and what Goldin and Katz (2024) refer to as anything that impacts income and productivity in the future. A fundamental underlying principle in human capital theory is that people have capaci-

ties that are comparable to other resources that can be used for the production of goods and services. Core to human capital theory is that investments in the capacities of people, their KSAOs, can be increased through investments in training and diverse forms of capacity building, and these investments are capable of yielding different forms of return (Schultz, 1961). Human capital theory emphasizes education, training, and the acquisition of knowledge in all forms to enhance creativity and the production capacities of people. Like other resources, when human capital is effectively utilized, it is capable of yielding benefits to the individual and any *organizational* context such as firms, societies, and countries, to which the individual belongs. Consequently, people and the human capital embodied in them are a form of capital for a country's development. Primarily, because human capital, which is basically the KSAOs individuals carry, is embodied in people, each person is the owner of his or her capital. Countries cannot actually own any human capital in the true sense of the word "own." However, countries can own a form of group-level human capital that emerges from the collection of people with KSAOs in particular areas.[1] National human capital can emerge, albeit through a complex process involving several factors, from the human capital of individuals, and become a collective form of human capital at country or national levels. Some countries are renowned for some occupations, skills, products, and services. For example, despite not having cocoa as a natural resource, Switzerland is famous for its quality chocolates and was once famous for watchmaking; in the world of sports and football, Brazil has become famous for their footballing skills. France is known for its fine wine and dining, and the United States (USA) has become associated with the Silicon Valley. Still, even such national human capital is not directly owned per se by the country as it ultimately has its roots in the KSAOs of individuals. Therefore, countries can put systems, formal institutions, and frameworks in place that enable individuals to acquire diverse KSAOs that have

---

[1] See Chaps. 2 and 3 of this book.

productive capacities from which a form of collective human capital at the national level can and could emerge.

Resource-Based Theory (RBT) According to RBT, a firm's growth, competitive advantage, and performance results from having resources that are valuable, rare, inimitable, and nontransferable and not necessarily from just its position in the product market (Barney, 1991; Penrose, 1959; Wernerfelt, 1984). Knowledge-based resources in particular should be developed by the firm internally as core competencies, while other resources can be outsourced (Prahalad & Hamel, 1990). With core competencies and capabilities developed by a firm from its resources, it can outcompete its competitors in the product market. RBT shifts the focus of competition among firms from product markets to resource markets or what is commonly referred to as factor markets. Core to the propositions of RBT is the heterogeneity of resources across firms. Although RBT is a preeminent theory of strategic management that is principally applied to firms, it is also relevant to the organizational context of countries. Arguably, RBT may even be more applicable to countries than it is currently to business firms because resource management is a core function in the governance of countries (see Table 10.1 for a summary of human capital theory, RBT, and transaction cost economics, and their implications for human capital accumulation in countries). As the fundamental resource of any country, people's human capital are valuable. The rarity, inimitability, and nontransferability of the human capital of a people in a country depends fundamentally on the country's strategic management of the resource. Countries in search of a strategic goal or vision such as attaining sustainable development must therefore shift focus to a core resource, human capital and its strategic management.

In reference to human capital, given the diversity of KSAOs in different individuals, each country has enormous amounts and potential for diverse forms of human capital. The strategic management of a country's human capital can result in attaining a national competitive advantage among counties. Although the goal of firms, especially business firms, and countries may differ substantially, countries continue to strive for economic growth and global competitiveness and political relevance. Interestingly, history has shown that one of the major determinants of political relevance is economic development and might. It is no coincidence that the countries with global political relevance are the countries with the global economic might. Even if countries do not deliberately aim for global political might, they strive for peace and economic prosperity, and strategic management of their human capital is the viable and most sustainable path toward attaining these goals.

Transaction Cost Economics The theory of transaction cost economics seeks to explain the optimum organizational structure that minimizes transaction costs. Transaction costs, or what can also be regarded as costs of exchange or costs of running the economic system of firms and the market mechanism (Williamson, 1975), include costs of coordinating, monitoring, controlling, and managing transactions. Following the logic of transaction cost economics, transactions can either take place within the governing structure of the market mechanism or within the firm, and the optimal governance structure for a transaction is the one that minimizes transaction costs (Coase, 1937; Williamson, 1975). Consequently, for any transaction such as resource acquisition, for example, firms have the three options to either make in-house within the firm, buy from the market, or a hybrid of make and buy in its various forms such as contracts, strategic alliances, and the like.

Clearly, the nature of transaction costs relating to countries' acquisition and accumulation of human capital differs from that of firms. However, some aspects of the logic of transaction cost economics can be applied to countries' accumulation of human capital. For countries, in developing or accumulating their human capital, they have to seek the path that minimizes all associated costs

Table 10.1 Theoretical framework for the generic approaches

Concept/theory	Major tenets	Implications for human capital accumulation
Human capital	Individuals have different KSAOs that can be human capital. Individuals own their human capital and can apply it in different organizational contexts including countries.	Countries do not own any human capital but can accumulate human capital by making investments in people and providing the systems and institutional frameworks required to facilitate people's acquisition of human capital.
Resource-based theory (RBT)	RBT emphasizes the relevance of resources that are valuable, rare, inimitable, and nontransferable. Knowledge-based resources are particularly relevant. Human capital fits well into a strategic resource.	Countries in search of sustainable development and a national competitive advantage should shift focus away from natural resources to the strategic management of human capital as its core resource.
Transaction cost economics	There are costs such as the costs of coordinating, monitoring, controlling, and managing transactions, within different governing structures of transactions. Transactions can take place within the firm, the market, or using a hybrid of the two. The optimum governing structure is that which minimizes transaction costs.	Countries' approach to human capital accumulation can either be internal, external, or a hybrid of the two. The optimum option is a combination of internal and any of the other two after consideration of economic, social, cultural, and other costs.

which is by no means limited to financial or economic costs. The path to human capital accumulation is either to accumulate human capital internally within the country, externally from outside the country, or through a hybrid combination of these two (Fig. 10.1). The most effective approach will be that which minimizes the various costs of human capital accumulation. The costs to be considered by countries go far and beyond financial and economic costs and include considerations of social and cultural costs of the path to human capital accumulation chosen.

The Three Generic Approaches to Countries' Human Capital Accumulation

In many ways, the three generic approaches countries can use for human capital accumulation are akin to a firm's strategic decision of its mode of employment of people with human capital. It is similar to what Lepak and Snell (1999), scholars of strategic human resource management, refer to as a firm's strategic human capital management decision of "make," internalize employment by building and developing the skills of employees, and "make and buy," externalize employment by outsourcing some functions to the market in addition to internalizing employment. Generally, in resource acquisition, the conventional decision for a firm is either to "make" in-house, within the firm, or to "buy," outsource from outside the firm. For a country, the two generic strategic approaches, of "make" and "make and buy," translate to internal human capital accumulation through human capital development and external human capital accumulation most notably through immigration. In contrast, unlike firms, the route of internal human capital accumulation through human capital development is not an option for countries (see Fig. 10.1). It is an obligation! Providing a functional education system is one of the fundamental responsibilities of governments in modern economics times. In addition to human capital development, countries may also opt for the decision to accumulate human capital externally either through immigration or by sending its people to acquire human capital from other countries or through

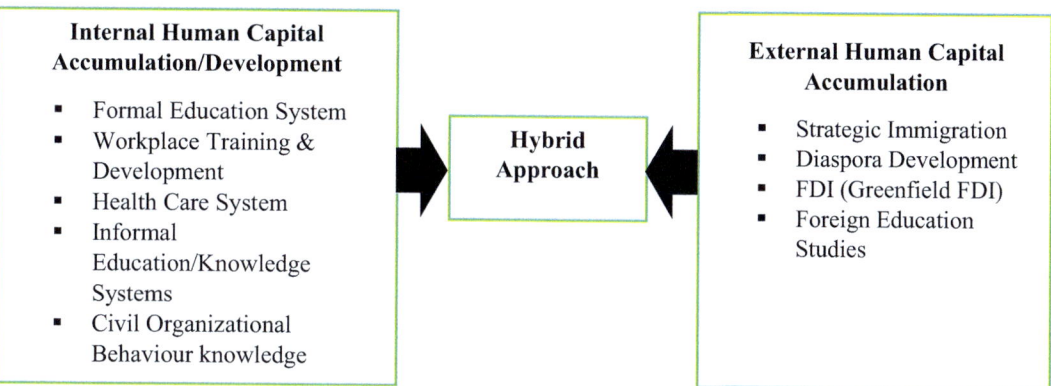

Fig. 10.1 Generic approaches to human capital accumulation for countries

foreign direct investments (FDI) especially greenfield FDI (see Fig. 10.1).

Path 1: Internal Human Capital Accumulation Through Human Capital Development

Human capital development is the process through which countries enhance the KSAOs of the people within their countries in order to increase their potential. All activities, processes, and systems that increase the productive capacities of people within a country's constitute human capital development. Consequently, all investments in these are human capital development investments. Investments in formal and informal education systems, institutions, and systems that enable and equip people to acquire and improve their KSAOs are investments in human capital development.

A fundamental component of a country's internal human capital development is its education system which may consist of pre-primary, primary, secondary, and tertiary education in various forms and combinations. Investments in health-care systems also constitute human capital development investments as they contribute toward directly enhancing the capacity of people to acquire the KSAOs that constitute human capital. Without good health, people's capacity to acquire and accumulate human capital is severely inhibited. Apart from education and health,

human capital can also be accumulated through work experience in the workplace (Jedwab et al., 2023; Lagakos et al., 2018). Although countries do not have as much direct impact on workplaces as they have with the education system, policies that encourage workplaces to train and develop their employees contribute to human capital accumulation and development. The case studies of Singapore and Finland presented in Chaps. 7 and 8, respectively, are examples of countries that have used this approach.

Path 2: External Human Capital Accumulation

External human capital accumulation consists of bringing in people with human capital into the country so they can apply their human capital to productive activities within the country. There are different routes through which a country may acquire and accumulate human capital externally. An obvious route is through immigration—attracting and bringing in people with human capital into the country. Historically, Canada whose case study is presented in Chap. 8 is a good example of a country that has made extensive use of this approach. Using strategic immigration as a human capital accumulation route also presents its own peculiar challenges which must be overcome for optimization. Proper integration of the immigrants with human capital into the economic and social systems of the country is key to reaping the benefits of using this method of external human capital accumulation. Language training, job search support, lower barriers to entrepreneurship, and other support need to be given to immigrants into the country for the country to effectively utilize their human capital. One of the foremost challenges of external human capital accumulation is actually attracting the people with the human capital a country requires for growth. This challenge has been exacerbated in recent times as countries silently compete for talent and human capital.

The Silent Global Competition for Talent and Human Capital

In 1998, the highly sought after global management consulting firm, McKinsey, released a report titled "the war for talent." The report highlighted the intense competition for recruiting and retaining talented employees, especially current and future top management executives, among companies. A subsequent report, a part two issue on the same subject with the same title published in 2001, also emphasized the intensified competition for talent among companies. These two reports bring attention to how companies are continuously engaging in an intense competition for the most talented workforce.

However, there seems to be a new ongoing competition. This new competition is silent. Few people seem to be taking notice or talking much about it, at least not in the language that takes cognizance of its competitive nature. Many appear to be taking no notice that it is a competition. It is a silent competition. It is the global competition for talent and human capital among countries. There is an ongoing but silent global competition for talent and human capital among countries. Countries, especially advanced countries going through what can rightly be called a second demographic transition, a shrinkage in the number of people in the traditional working age population, are in a silent war to attract the best talent and human capital to augment their aging population and slowing birth rates. The deliberate immigration programs to attract the immigra-

tion of skilled and highly skilled workers from other countries constitute a form of competition among countries for human capital accumulation. Skilled migration, especially from the global south to the global north, has been steadily increasing over the years (Bailey & Mulder, 2017; Boucher & Cerna, 2014; OECD, World Bank, & ILO, 2015), and much of it has been a deliberate bid to attract human capital.

This new silent competition is not just among developed countries. There is an aspect of it that is also being played out between developed and developing countries. The competition presents a different facet for each of these two sets of countries. For developed countries, the competition is to fill job gaps that require a skilled workforce that cannot be readily filled from within their countries. Countries such as Australia, Canada, Germany, United States (USA), United Kingdom (UK), and other advanced countries have put in place immigration programs that encourage certain skilled migrants to come into their countries to fill the shortages of the workforce. Germany, for instance, recently introduced "Chancenkarte," a work visa program that offers opportunities through a point system to non-European Union citizens to find work opportunities and migrate to Germany. Some of these countries use a point-based system that gives points for factors such as the potential immigrant's age, educational attainment, language proficiency, and occupation or profession. These immigration programs are targeted at specific occupations and professions and are aimed at ensuring that the country continues to have the necessary human capital required for certain jobs and ultimately to ensure that their economies continue to grow.

Contrary to some popular media, developed countries, and indeed some countries, are not necessarily averse to immigration. It's the type of immigration and the nature of immigrants that really matter. Controlled, and strategic, immigration presents opportunities for host countries to accumulate human capital and the potential benefits that can be derived from access to people with such requisite human capital. Skilled migrants have become critical to the growth of developed and mature economies (Bailey &

Mulder, 2017; Tharenou & Kulik, 2020). To avoid a possible stagnant economic growth, many developed and matured economies are becoming more open to immigration but immigration of skilled workers.

For developing countries, such as most African countries, the face of the competition is the same except that it adds another layer which presents the competition as if it is in an opposite direction to that being faced by developed countries. The global competition for talent and human capital presents the challenge of brain drain for these countries. Developing countries are striving to keep a skilled workforce from leaving their home countries to greener pastures which is most often times to developed countries. Recently, the International Monetary Fund (IMF) estimated the expected number of African immigrants to Organisation for Economic Co-operation and Development (OECD) countries by 2050 to rise to 34 million (IMF, 2016; Kweitsu, 2018). Consequently, many of the "brain drain" of African countries result in the skilled immigrants of developed countries. In a sense, one country's loss becomes the gain of another country. What one country loses in terms of human capital becomes the gain of another country. Interestingly, some sectors such as the health and education sectors, the two key sectors directly involved in human capital development, have been particularly hit hard by the emigration of skilled workforce outside their home countries (Ebeye & Lee, 2023; Kalipeni et al., 2012). According to available data from the World Health Organization (WHO), on the average, African countries have less than 0.5 doctors per 1000 people (Kweitsu, 2018; WHO, 2020).

The brain drain, perhaps more aptly the emigration of human capital, from African countries to other countries is having some adverse effects on some sectors of the economy. The Global Code of Practice on the international recruitment of health personnel instituted by the World Health Organization in 2010 was an attempt at stemming the negative effects of the migration of skilled health personnel on the health sector of developing countries (Bailey & Mulder, 2017; Taylor et al., 2011). There is a dire need to ensure that

this is not merely seen as a brain drain, or skilled migration as it is often referred to in the migration literature, but as one that is a form of competition, though a silent one, among countries for human capital. Only then can countries, especially African countries, take seriously the necessary actions to strategically develop their human capital and to put in place policies and environments that encourage Africans with human capital to stay in Africa.

The exact nature of the economic impacts of skilled migrants to developed host countries and developing home or origin countries, however, is still an issue of debate (Bailey & Mulder, 2017; Nathan, 2014). Skilled migration can increase innovation and entrepreneurship and even alter FDI flows in both destination and origin countries. Apart from the adverse impacts on certain sectors in developing origin countries such as its impact on the health sector mentioned above, some point to the remittances, knowledge transfer, and other economic benefits that migrants potentially bring to their home or origin countries. It is safe to conclude that the summary impacts on skilled migration on both developed and developing countries depend on a number of factors. However, putting skilled migration within the human capital discuss, as a competition between countries for human capital, presents another perspective for countries, especially developing countries such as those within Africa. It appears that developed countries are much more aware of the competition for human capital that skilled migration represents than developing countries.

Regardless, each country faces a different side of the same competition. Some major pull factors are driving this competition in developed countries, while some push factors are the major drivers in developing countries. A major pull factor is the aging demographics in developing countries. Developed and matured economies are going through a demographic transition that is seeing an increase in the numbers of the aging population, coupled with low and reducing birth rates. On the average, the median age for developed countries is 40 years or above compared to 20 years for African countries (Ritchie & Roser,

2019; OECD, 2019; UN, 2022). Developed countries also have a higher life expectancy than their developing counterparts, and birth rate per woman continues to decline. The average life expectancy in Europe is 77 years compared to 61.7 years in Africa, for example (Dattani et al., 2023; UN, 2022). An aging population and reducing birth rates translate to a reducing population within the working age group, increasing dependency ratios, and decreasing taxable persons and incomes. An OECD report in 2019 categorically stated that "without policy action, growing number of retirees will strain public budgets and slow economic growth" (OECD, 2019). One public policy direction of some matured economies is policies and programs that attract skilled workers. The aim is to mitigate the adverse effects of aging populations and reducing birth rates. On the opposite side, a listing of the push factors in many origin developing countries almost always includes a perception of dwindling economic opportunities.

Countries can win this war for talent and human capital. Indeed, African countries can win this silent war for human capital, but they must see investments in human capital and the strategic management of their human capital resources as national priorities. Human capital accumulation through deliberate human capital development must become a priority for African countries. It is no longer enough to simply allow education and health systems, the core components of the human capital development system for countries, to go on operating normally without any strategic objectives and direction. Strategic investments on developing human capital internally within the country need to be made. For countries to reap the benefits of investments in human capital development, the persons in which the human capital is embodied have to be willing to deploy the human capital in the country. Consequently, encouraging people with human capital to stay in their countries of human capital development needs to go alongside deliberate strategic human capital investments and management.

Other Forms of External Human Capital Accumulation It is well established that FDI can transfer knowledge and technology from foreign companies and countries to the domestic economy (Paul & Feliciano-Cestero, 2021). What is less obvious and often less discussed and researched is that FDI also impacts human capital development in the recipient country (Emako et al., 2023) and can serve as a means of transferring human capital across countries. Consequently, countries have the option of using FDI to not only acquire physical and other forms of capital but to also acquire and accumulate human capital. Inward greenfield FDI in which parent companies establish their presence in other countries can especially help countries to accumulate human capital as people gain work experience and undergo training and development in the established multinational companies.

Another route through which countries can accumulate human capital is foreign education of its citizenry. The human capital acquired by international students invariably are a form of human capital accumulation when such students return to their home countries. Countries can strategically use this route to accumulate human capital by putting programs and policies in place that encourage their nationals to undertake formal studies in foreign countries known for quality education systems and the return of such students. China has strategically used this to accumulate human capital for technology transfer, FDI, entrepreneurship, and innovation and to create global social network ties through *Haigui*[2] (Kun, 2011; Lefifi & Kiala, 2021). However, the possibility of students flouting a return to their countries on completion of their studies is a very present risk that countries that go through this route face. Nevertheless, it's a route that is open, especially to African countries. The onus lies in designing a system that is both beneficial to the student and home country and one that encour-

---

[2]Translates to "overseas returnees" in English and is an informal colloquial metaphor in China used to refer to Chinese students who returned home after studying abroad (Lefifi & Kiala, 2021).

ages return upon completion of studies. African countries could make use of the Human Capital Returns Matrix as a guide to designing programs, policies, and systems that encourage students that go abroad for foreign education to return to their home countries.

The Human Capital Returns (HCR) Matrix

The HCR matrix presented in Fig. 10.2 assists in knowing whether or not to make investments in the foreign study of their students and generally the amounts of investments to make. The key for home sponsoring countries is knowing the right amounts of financial investments that should be made in support of students travelling abroad to acquire human capital and to return to their home countries. The numbers in the quadrants are the order of priority in which countries should consider when making investments in foreign education of their nationals. Clearly, the optimal quadrant that increases the chances of students returning to their home countries and one that simultaneously minimizes loss of funds is quadrant 1, where the human capital being acquired has high benefits to both the individual and the country. The matrix has general utility in analyz-

ing when to make private human capital investments in individuals regardless of whether the individual is studying within the home country or abroad in a foreign country.

A precedence to using the matrix is that countries have evaluated the areas of their current and potential human capital needs by conducting a human capital or skills audit. Indeed, one of the first steps in a country's strategic human capital plan is to conduct a human capital audit of the types and levels of knowledge, skills, and abilities that is available in the country and to the country. Without this initial step, actions taken toward accumulating human capital may not be effective. Moreover, what can be regarded as human capital changes with time. Therefore, there is the need to ensure that the country has the stock of human capital that will continue to be relevant.

Moreover, it is difficult to assess progress with the strategic plan without knowledge of the starting point. A human capital audit enables countries to assess their stock of human capital and can assist in putting a national strategic human capital plan in place. The idea, for example, of offering full scholarships for every course of study in foreign countries without recourse to the specific benefits for the country is an inefficient and negligent system of human capital development and accumulation. Conducting a human capital audit that consists of evaluating the human capital profile and stock of human capital available to the country is a worthy exercise that should precede and be used in combination with the human capital returns matrix.

Finally, diaspora development is a viable way of accessing human capital externally for countries, but this route often lacks formal coordination. Available data shows that remittances from the diaspora form a substantial part of foreign inflows of low- and middle-income countries and was estimated to have reached $669 billion for countries in this category in 2023 (World Bank, 2023). According to the Global Knowledge Partnership on Migration and Development (World Bank-KNOMAD December, 2023; see Fig. 10.3), remittances into Africa have hovered around a hundred billion US dollars (US$100 bil-

		Low	High
Private Returns/Benefits to Individual	High	3 Investments depends on other considerations	1 Full Investments
	Low	4 No Investments	2 Partial Investments
		Low High	
		Public Returns/ Benefits to Country	

Fig. 10.2 The Human Capital Returns (HCR) Matrix

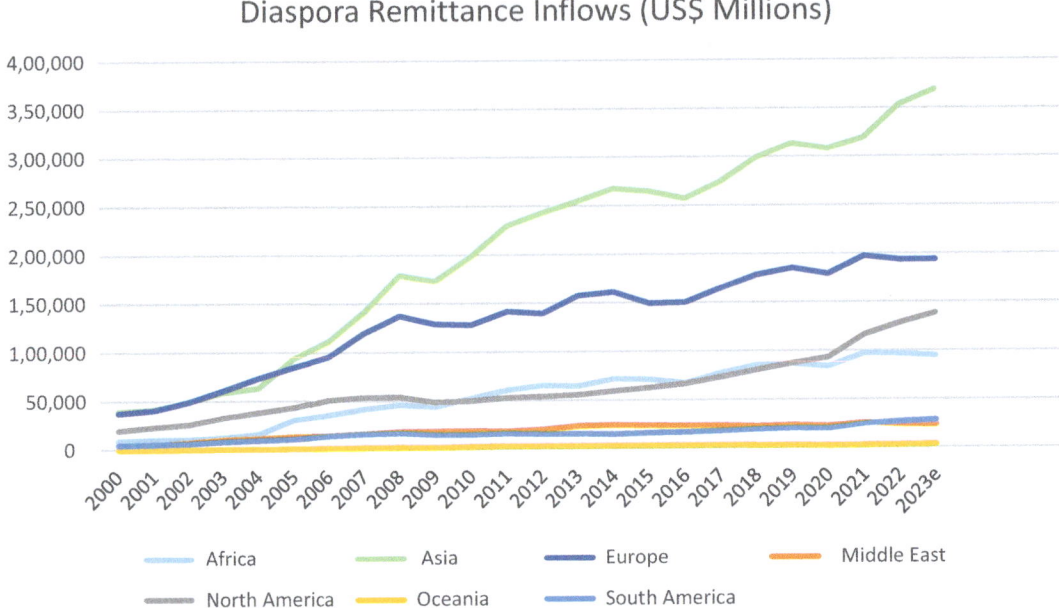

Fig. 10.3 Diaspora remittance inflows by geographical continental regions. (Source of data for graph: World Bank-KNOMAD December (2023))

lion) in recent years, up from less than 10 billion dollars in the year 2000 (US$10 billion). This is a signal of how Africans in diaspora have increased in recent times and which is also as a result of the increasing brain drain discussed earlier in this chapter. Apart from remittances, the diaspora can be a real source of human capital to countries. The onus is on countries to develop the formal means of harnessing the knowledge, skills, abilities, and other characteristics of individual nationals in the diaspora for their benefit.

Path 3: The Hybrid Approach—The Practical Approach

In reality, no country can depend on accumulating human capital solely internally or externally. Other than internal human capital development and accumulating human capital externally, the practical approach to human capital development is the hybrid approach. This entails a combination of both approaches wherein a country in addition to developing its human capital inter-

nally also combines this with acquiring human capital externally from outside the country. Elements of a hybrid approach can consist of formal education system, a sine qua non for every country, combined with some other elements from the other two generic approaches in a carefully thought-out strategic plan. The three case studies presented in this book, Singapore, Finland, and Canada, are in reality hybrid approaches. Although Canada has leaned toward using external accumulation of human capital, there has been no neglect of developing human capital internally. Finland and Canada also have had some elements of external human capital accumulation. Although countries may lean toward the use of one approach or the other, in reality, the hybrid approach is the practical approach.

These three generic approaches discussed above somewhat align with the five main categories of human capital identified in early studies of the concept during the conceptualization stage: health, on-the-job training, schooling, adult education, and migration (Schultz, 1961; Teixeira,

2014). These five categories are elements of the generic approaches presented. A country is capable of accumulating human capital through investment in these areas and other elements in Fig. 10.1.

Avoiding the Micro–Macro Paradox

Whatever approach and routes to human capital accumulation chosen, the onus lies with countries to ensure that they avoid the "micro–macro paradox," the situation where more human capital accumulation, especially in terms of formal schooling, only benefits the individuals and not the country (this is explained more fully in Chap. 4, the chapter on human capital outcomes). One way of avoiding this paradox is to provide an enabling environment for business, investments, and entrepreneurships to strive such that human capital can be effectively deployed into production enabling activities and away from rent seeking. Governments must also ensure that they are not the vast employer in the economy and that direct investments in private education of persons such as scholarships are made considering the Human Capital Returns Matrix.

Conclusion

Each of the three generic approaches has their merits and demerits. However, as has been noted, it seems rather obvious that the hybrid approach, in which human capital is developed internally and also accumulated externally from sources outside the country, is the pragmatic approach for countries. The hybrid approach is also simultaneously a two-pronged approach, combining elements from the other two approaches. Regardless of the approach adopted, in the midst of a silent global competition for talent and human capital, it has become imperative for African countries to put systems in place for the strategic development and accumulation of human capital.

References

Bailey, A., & Mulder, C. H. (2017). Highly skilled migration between the Global North and South: Gender, life courses and institutions. *Journal of Ethnic and Migration Studies, 43*(16), 2689–2703. https://doi.org/10.1080/1369183X.2017.1314594

Barney, J. (1991). Firm resources and sustained competitive advantage. *Journal of Management, 17*(1), 99–120.

Becker, G. S. (1964). *Human capital: A theoretical and empirical analysis, with special reference to education.* University of Chicago Press.

Becker, G. S. (1993). *Human capital.* University of Chicago Press.

Boucher, A., & Cerna, L. (2014). Current policy trends in skilled immigration policy. *International Migration, 52*(3), 21–25. https://doi.org/10.1111/imig.12152

Coase, R. H. (1937). The nature of the firm. *Economica, 4*, 386–405.

Dattani, S., Rodés-Guirao, L., Ritchie, H., Ortiz-Ospina, E., & Roser, M. (2023). *Life expectancy.* Published online at OurWorldInData.org. Available at https://ourworldindata.org/life-expectancy

Ebeye, T., & Lee, H. (2023). Down the brain drain: A rapid review exploring physician emigration from West Africa. *Global Health Research & Policy, 8*, 23. https://doi.org/10.1186/s41256-023-00307-0

Emako, E., Nuru, S., & Menza, M. (2023). The effect of foreign direct investment on capital accumulation in developing countries. *Development Studies Research, 10*, 1. https://doi.org/10.1080/21665095.2023.2220580

Goldin, C., & Katz, L. F. (2024). The incubator of human capital: The NBER and the rise of the human capital paradigm. In *The economic history of American inequality: New evidence and perspectives.* National Bureau of Economic Research.

IMF. (2016). *World economic outlook: Subdued demand: Symptoms and remedies.*

Jedwab, R., Romer, P., Islam, A. M., & Samaniego, R. (2023). Human capital accumulation at work: Estimates for the world and implications for development. *American Economic Journal: Macroeconomics, 15*(3), 191–223.

Kalipeni, E., Semu, L. L., & Mbilizi, M. A. (2012). The brain drain of health care professionals from sub-Saharan Africa: A geographic perspective. *Progress in Development Studies, 12*(2–3), 153–171. https://doi.org/10.1177/146499341101200305

Kun, C. (2011). *Producing China's innovative entrepreneurship: Nationalism, cultural practices, and subject-making of transnational Chinese Professionals.* University of California, Berkeley, ProQuest ID: Chen_berkeley_0028E_11567. Merritt ID: ark:/13030/m58919x5. Retrieved from https://escholarship.org/uc/item/4qn2m66v

Kweitsu, R. (2018). *Brain drain: A bane to Africa's potential*. Available at https://mo.ibrahim.foundation/news/2018/brain-drain-bane-africas-potential

Lagakos, D., Moll, B., Porzio, T., Nancy Qian, N., & Schoellman, T. (2018). Life-cycle human capital accumulation across countries: Lessons from US immigrants. *Journal of Human Capital, 12*(2), 305–342.

Lefifi, T., & Kiala, C. (2021). Untapping FOCAC higher education scholarships for Africa's human capital development: Lessons from haigui. *China International Strategy Review, 3,* 177–198. https://doi.org/10.1007/s42533-021-00074-y

Lepak, D. P., & Snell, S. A. (1999). The human resource architecture: Toward a theory of human capital allocation and development. *Academy of Management Review, 24*(1), 31–48.

Nathan, M. (2014). The wider economic impacts of high-skilled migrants: A survey of the literature for receiving countries. *IZA Journal of Migration and Development, 3*(1), 1–20.

OECD. (2019). *The challenge of aging: Working better with age*. Available at https://www.oecd-ilibrary.org/sites/d56a2fbc-en/index.html?itemId=/content/component/d56a2fbc-en

OECD, World Bank and ILO. (2015). *The contribution of labour mobility to economic growth*. OECD.

Paul, J., & Feliciano-Cestero, M. M. (2021). Five decades of research on foreign direct investment by MNEs: An overview and research agenda. *Journal of Business Research, 124,* 800–812. https://doi.org/10.1016/j.jbusres.2020.04.017

Penrose, E. T. (1959). *The theory of the growth of the firm*. Wiley.

Prahalad, C. K., & Hamel, G. (1990). The core competence of the corporation. *Harvard Business Review, 68*(May–June), 79–91.

Ritchie, H., & Roser, M. (2019). *Age structure. Our world in data*. Available at https://ourworldindata.org/age-structure

Schultz, T. W. (1961). Investment in human capital. *The American Economic Review, 51*(1), 1–17.

Taylor, A. L., Hwenda, L., Larsen, B. I., & Daulaire, N. (2011). Stemming the brain drain – A WHO global code of practice on international recruitment of health personnel. *New England Journal of Medicine, 365*(25), 2348–2351.

Teixeira, P. N. (2014). Gary Becker's early work on human capital – Collaborations and distinctiveness. *IZA Journal of Labor Economics, 3,* 12. https://doi.org/10.1186/s40172-014-0012-2

Tharenou, P., & Kulik, C. T. (2020). Skilled migrants employed in developed, mature economies: From newcomers to organizational insiders. *Journal of Management, 46*(6), 1156–1181. https://doi.org/10.1177/0149206320921229

UN. (2022). *World population prospects*. Available at https://population.un.org/wpp/Download/Standard/MostUsed/

Wernerfelt, B. (1984). A resource-based view of the firm. *Strategic Management Journal, 5*(2), 171–180.

WHO. (2020). *Physicians (per 1,000 people)*. Available at https://data.worldbank.org/indicator/SH.MED.PHYS.ZS

Williamson, O. E. (1975). *Markets and hierarchies: Analysis and antitrust implications*. Free Press.

World Bank. (2023). *Remittance flows continue to grow in 2023 albeit at slower pace*. Press release, Press Release No: 2024/040/SPJ, December 2023. Available at https://www.worldbank.org/en/news/press-release/2023/12/18/remittance-flows-grow-2023-slower-pace-migration-development-brief#:~:text=%E2%80%9CRemittance%20flows%20to%20developing%20countries,strengthening%20a%20country's%20debt%20position

World Bank-KNOMAD December. (2023). *Migration and development brief 39*. Available at https://knomad.org/publication/migration-and-development-brief-39

Developing Africa's Human Capital: Recommendations and Suggestions

11

Abstract

This chapter presents some recommendations and suggestions on how African countries can develop and accumulate human capital. The recommendations go beyond the broad generic approaches identified in the previous chapter to more specific ones that can help African countries to formulate programs and policies on human capital development and accumulation. The recommendations are embodied with in-depth theory-based explanatory notes that expound the rationale behind each proposed guideline to different groups of stakeholders in Africa's sustainable development.

Keywords

Education system · Experiential learning · M-STEM · Immigration · Human capital

Introduction

With a population of over 1.4 billion people and the continental region with the youngest population, African countries have invaluable resources beyond their natural resources. Of the 1.4 billion people on the continent, about half (50%) of them are below 19 years old. This translates to a population that consists of an estimated 700 million youths that are below the age of 19. According to projections by the United Nations (2022), it is expected that there will be 2.5 billion people in Africa by 2050, which will be about 25% of the world's population. Half of this projected population figure, the median age, will be youths below 25 years old (UN, 2022). By the turn of the century in 2100, Africa is projected to have a population of 4 billion, two out of every five people on earth are expected to be Africans, and the median age in Africa is expected to be 35 years (Stanley, 2023; UN, 2022). This means that by 2100, 2 billion Africans are expected to be below 35 years. Africa, now and in the near future, is not just experiencing a population boom. Africa is undergoing a youthful population boom.

Concealed within these data and statistics is the huge potential for human capital and economic growth that could result from demographic dividends among numerous other benefits. African countries have a large youthful population with massive potentials for harnessing human capital. While the world and Africans themselves seem to be excited about the abundant natural resources in African countries, the true incubators of the continent's wealth are the people: the youthful population. The urgent need for African countries to make human capital development, and the accumulation of human capital, a strategic priority cannot be overemphasized. This chapter discusses some recommendations that can be adopted and adapted to ensure that the continent's most valuable resources are developed. Unlike the generic strategic approaches discussed in the previous chapter, this chapter delves into more specific recommendations and suggestions that can help African countries to develop and accumulate human capital.

Human Capital Planning, Audit, and Tracking

The first recommendation is the need for African countries to have a strategic human capital development plan. A prerequisite to this is taking general stock of the human capital available to the country by conducting a human capital audit. The essence of a human capital audit is for countries to have knowledge of the skills base of their population and workforce in order to be able to assess gaps, to aid strategic planning and human capital investments, and to guide policy-making on human capital. As a major component of human capital, a human capital audit must of necessity include an audit of the health of the population to assess the overall health of the population. Subsequent to an initial human capital audit, a periodic tracking of the country's stock of human capital can be done on a continuous basis to continue the assessment of the state of the country's human capital. Hence, a human capital audit is not a one-time activity. It is one that should be conducted periodically as deemed fit by each country. This will aid monitoring of the progress and outcomes of any specific human capital development and accumulation policies and programs.

Various global human capital indexes available such as World Bank's Human Capital Index (HCI), the World Economic Forum (WEF)'s Global Human Capital Index (GHCI), United Nations' Human Development Index (HDI), and other databases of national human capital measures[1] are all attempts at measuring and taking account of countries' stock of human capital. Despite the concerns of some about the method-

ology and validity of some global human capital indexes, they can still be used in conjunction with the results of a human capital audit. The outputs of human capital audits can also be used jointly with any of the global human capital indexes that is perceived as more relevant and appropriate to assess the state of a country's stock of human capital. These indexes can serve as good starting points to conducting a human capital audit for countries. However, these indexes may not aptly capture the relevant human capital components for a country to make concise and decisive policy-making decisions, hence the need for a human capital audit that directly suits the needs and requirements of particular countries.

Effective human capital audits need to incorporate assessments and measurements of quality and not just quantity of human capital. Quality human capital is a necessity for economic growth and not quantity alone (Aziegbe-Esho & Anetor, 2020; Hanushek & Woessmann, 2008; King & Winthrop, 2015). However, measures of countries' human capital, including the global human capital indexes, are more readily amenable for measuring quantity, rather than quality of human capital. Still, metrics for measuring quality of human capital, which lean toward actual learning outcomes and health outcomes, have to be incorporated into national audits for it to be an effective measure and a reliable guide to policy-making. Unfortunately, it can be incredibly complex and difficult to measure the quality of education and by extension the quality of human capital. Thankfully, health is simpler to assess. The ratio of students to teachers, class size, availability of textbooks and instruction materials, and learning environment all contribute to education quality (GEM, 2022), so do the quality of teachers and teachers' training. To assess quality and inculcate quality into the human capital audit, programs that measure scores from standardized learning assessments conducted for students at various levels of education could be initiated at national and regional levels. These kinds of test assessment scores[2] measure quality and should form part of a country's human capital audit.

---

[1] Chapter 5 of this book presents a more comprehensive description of the various available global human capital indexes.

[2] OECD's Programme for International Student

The human capital audit should precede a national human capital strategy. Other components of a national human capital strategy are largely dependent on the outcomes of the human capital audit and on the particular national goals. Ideally, the national strategic human capital plan should form part of a country's national strategic developmental plan. An audit and tracking of a country's human capital, when done properly and periodically, enables countries to take stock of the nature of human capital currently available. It also enables countries to compare the nature of their current stock of human capital with the human capital required to attain countries' strategic objectives. For example, given the potential for foreign direct investment (FDI) to boost economic growth (Alfaro et al., 2000; Anetor et al., 2020; Adams, 2009), many African countries continue to take initiatives to attract FDI into their countries (Njuguna & Nnadozie, 2022). However, for FDI to impact economic growth, it is not enough to simply attract FDI into a country. As the case of Singapore that was explored in Chap. 6 highlights, FDI impacts growth much more when it is aligned with the human capital stock already existing within the country. This is also in consensus with recent extant studies and research findings on FDI and economic growth. All FDI are not equal because their effects and impact are not equal. The impact of FDI on recipient countries depends on many complementary factors one of which is the nature of human capital available in the country (Bénétrix et al., 2023; Bruno et al., 2018; Su & Liu, 2016). Consequently, African countries looking for FDI and envisaging its contribution to economic growth need to be aware of the exact nature of human capital available in their countries. Conducting a human capital audit and continually tracking the nature of human capital in the country, as well as ensuring that this is done periodically, are prerequisites for a national strategy focused on FDI-driven eco-

nomic growth. Human capital audits arm countries with the -knowledge of the right FDI to pursue and attract. Otherwise, the FDI attracted might not make the expected economic impact.

In the rest of the sections in this chapter, additional specific recommendations on human capital development and accumulation that can be adopted or adapted to suit particular needs of countries are presented.

Education: A Human Capital Formative Education System

Without doubt, education is paramount to human capital and its development. It is almost impossible to talk about human capital, and its development, without making reference to formal education systems. Gary Becker's foremost analysis of human capital in his books makes reference to education, particularly formal education, as a core investment in people and human capital. Consequently, the formal education system at all three levels of education, basic or primary level, secondary school education, and tertiary of higher education levels, are all important to human capital. However, unfortunately, the formal education system in many countries, not just African countries, has begun to revel in the creation and awarding of certificates and not necessarily in ensuring the formation of human capital. What is the difference? The major difference lies in whether the formal education process that culminates in the award of certificates has actually created or enabled the creation of the knowledge, skills, abilities, and other characteristics (KSAOs) that have economic producing capacity. Human capital is the KSAOs of individuals that can be put to productive use and have economic value. The education that leads to the award of certificates without imparting the KSAOs that can be put to productive use offers little to the individual owner and the organizational context and society to which the individual person belongs. In other words, education systems in African countries need to go beyond schooling to ensuring that learning and actual human capital formation

---

Assessment (PISA) and the Laboratorio Latinoamericano de Evaluación de la Calidad de la Educación (Latin-American Laboratory for Assessment of the Quality of Education (LLECE) in Latin American countrie, are examples of assessment tests that measure quality of learning.

takes place during the years spent within the formal education system.

From the data and statistics on Africa's huge youthful population, the point is often made of the huge potential of the large working age population of those between 25 and 64 years. By 2100, the number of people in the working age group, between 25 and 64 years old, will be about 1.5 billion (UN, 2022; World Bank, 2023a). The much-envisaged demographic dividends and the resultant economic growth will purportedly result because of the vast number of people within the working age population. While these potentials are a possibility, it is also often implicitly either forgotten, neglected, or ignored that the jobs for this enormous working age population need to be created. World institutions and international agencies such as the World Bank and African Development Bank have widely acknowledged that job creation for Africa's growing huge working age population should be the overarching aim of all policies and reforms. However, how will and should the jobs be created? Who will create the jobs? These are questions that need urgent workable answers. Meanwhile, some of the countries with the highest unemployment rates are African countries. The unemployment rates, which are already high, also mask a huge underemployment problem which is also sometimes hidden within the huge informal sector in many Africa countries.

Entrepreneurship, Entrepreneurial Problem-Solving Education, and Experiential Learning

The education that African countries need, and must have, is the education that is capable of truly creating human capital. It is one with the capacity to enable innovation, creativity, entrepreneurial thinking, and entrepreneurship and not just certificate-producing education of elites who see formal education only as a means of gaining employment in an already created workplace. Africa needs the kind of education that truly creates human capital that can create workplaces and not just mere certificates. Consequently, a

formal education that emphasizes entrepreneurship and entrepreneurial education needs to be promoted.

Much of the formal education, at all three levels, as is known today, especially tertiary education was developed to prepare students for the world of work (Carter et al., 2021; Casillas et al., 2019; Chamorro-Premuzic & Frankiewicz, 2019; Panth & Maclean, 2020). The pursuit of tertiary education particularly university education solely for the sake of "intellectual endeavors" has since almost come to an end. Much of the discuss on students' formal education is now on how to prepare students for the "world of work" to prevent a mismatch of the skills and competencies needed in the workplace and those taught in schools. However, many easily forget that this "world of work" needs to be created, particularly in contexts such as African countries where unemployment levels have traditionally been high. When coupled with the burgeoning youthful population, the need for a different form of formal education that can not only prepare students for the "world of work" but also prepare them with the capacity of creating the "world of work" becomes imperative. Moreover, thousands of graduates of higher educational institutions are finding it difficult to get jobs, while employers are at the same time complaining of unemployable graduates (Archer & Chetty, 2013; Mgaiwa, 2021; Obor & Kayode, 2022; Siivonen et al., 2023). Although graduates' unemployability phenomenon is not limited to African countries (Cheng et al., 2022; Grosemans et al., 2023; Siivonen et al., 2023), it is more pertinent in Africa given the context of a burgeoning youth population amid developmental and other challenges.

In many African countries, the "world of work" during and immediately after colonial era was largely limited to public service. Private enterprise as a means of work post-formal education was few and far between. This does not however mean that entrepreneurship, business, and enterprise was lacking in African countries (See Verhoef, 2017). However, the format in which they existed was not in the formal organization into firms and corporations, as it is today (Verhoef, 2017). Formal education in and immediately after colonial Africa was a means of gaining employment into government agencies and enterprises and in already established formal corporations and organizations in the "not so big private sector." The cumulative effect is that as time has grown, the public service has become too small to accommodate all graduating students of formal education. The foreign multinational companies have also not been enough to employ all graduating students. As noted by Goldin and Katz (2008), as enrolment in schools increases, the rate of return to education reduces. This seems to have been proven true in many African countries. As the number of students graduating and entering the workforce has increased, the workplaces available have become too few to accommodate the graduating students. The result is unemployment and reducing rate of private returns to education because of basic supply and demand. "There is a race between the supply of skills and the demand for skills with the return to education as the equilibrium price" (Goldin, 2016, p. 77). This aptly explains the situation in many African countries and also explains the low wages and compensation in comparison to what is obtained outside the continent, especially in advanced countries. One of the solutions lies in increasing the supply of workplaces. Otherwise, graduating students will emigrate to workplaces in geographical locations outside the continent where their skills are in great demand due to shortages in the labor market of those countries. Unfortunately, not much seem to have been deliberately done to change this narrative. If the current situation in many African countries continue, the continent will continue to lose people with human capital to other countries and continents and lose the silent global competition for talent and human capital.

Tertiary education in Africa has to rethink its role in the broad society beyond just preparing students for jobs and the current and future workplace. Higher education students need the capacity to not only work in the workplace but to also be creators of workplaces. African countries need to create jobs and this will involve creating companies. Therefore, formal education systems cannot afford to blindly follow the format of that of developed countries. There is a need to realize that entrepreneurship is paramount to creating jobs for African countries. Developed countries can afford to just develop human capital as there are already many organizations in existence to absolve the newly trained workforce. However, this is not the case in many African countries. Most African countries still do not have enough organizations and workplaces to absorb all the people that may be trained/educated. Also, it is doubtful that there can be enough foreign investments into African countries that will be enough to create the workplaces that are direly needed to absolve Africa's growing and teeming youthful population. Therefore, human capital development must be in conjunction with concrete entrepreneurial training accompanied with visible goals and results of creation of organizations, jobs, and opportunities for people to strive beyond just searching for "unavailable" employment.

Consequently, the education that is needed in African countries is one that incorporates entrepreneurship and one that is entrepreneurial. However, this goes beyond simply teaching entrepreneurship as a course or entrepreneurship education in higher education, even though this is a commendable starting point and many tertiary education institutions in Africa are already doing this. In Nigeria, for example, the national university commission has introduced and made entrepreneurship education compulsory in the curriculum of Nigerian universities (Igbokwe-Ibeto et al., 2018; Olutuase et al., 2018, 2023; Uzoka, 2006). Indeed, entrepreneurship education is one that should be taught at all three tiers of formal education: primary, secondary, and tertiary. However, entrepreneurial education will have to also entail inculcating "entrepreneurial problem-solving" into every course of study at higher education levels. This approach involves bringing specific societal problems relating to particular areas into the relevant subject areas both in the development of the curriculum and in the teaching. African countries need education systems that inculcate an entrepreneurial problem-solving mindset into students.

The type of entrepreneurship and entrepreneurial education that is being referred to here does not have its focus solely on profit-making. Its focus is not on creating "capitalist" ventures that aim to make profits for the sake of making profits. The focus of entrepreneurial problem-solving education is providing solutions to societal problems, particularly society's developmental problems and challenges, as it relates to specific areas of study. Therefore, the curriculum itself has to be designed and taught in a way that takes cognizance of society's challenges as it relates to different areas of study.

The current practice in some higher education institutions of teaching students to engage in various arts and crafts activities as part of their entrepreneurship education is largely deficient and insufficient. At best, such training endeavors are only equipping students with another set of handicraft skills that could truly become useful in earning a living outside of the primary course of study upon graduation. Rather, the entrepreneurial problem-solving education that is being advocated for here inculcates the capacity and attitudes of societal problem-solving through creating viable enterprises into the students. The aim is for students to creatively and innovatively think through societal problems relating to their course of study as they undergo their study such that they are able to create viable solutions to societal problems. They, therefore, may not need to delve outside of their areas of study to become entrepreneurs upon graduation or resort to engaging in ventures and creating enterprises that do not require formal higher education to establish. This type of education will involve rethinking the current "curriculum" in many African countries and how it is taught.

Perhaps, rethinking the current form of higher education is not relevant only for African countries but all countries. In his writing on technological unemployment that might result from the new digital technology age or what some refer to as the Fourth Industrial Revolution (4IR), Peters (2017) asks the question: "where will new jobs come from and what is the purpose for education especially at advanced levels when the covenant between higher education and jobs has been permanently broken?" (p. 4). This is a pertinent question. The new digital technology age, or 4IR, is characterized by accelerating and merging different technologies, notably artificial intelligence and the like. Rethinking the curriculum in higher education and incorporating entrepreneurial problem-solving components should essentially also take cognizance of the application of technology—digital and all related technologies. In its current form, higher education is playing catch-up to technological innovations and advancement. Generative artificial intelligence (AI) has once again shown that higher education needs a rethinking, not only in African countries but worldwide. Yes, clearly, equipping students for jobs should still be a critical role for higher education. Other critical roles clearly have to emerge, especially in the age of new and emerging technologies. For the continent of Africa, this is even more pertinent. One can argue that higher education has not really been able to proffer adequate solutions to the developmental challenges of African countries. Like it was stated, this fact is open for argument. Nevertheless, African countries cannot afford not to rethink higher education and its roles, especially considering their developmental and other challenges.

The form of education that is capable of inducing an entrepreneurial problem-solving mindset is one that transcends just receiving explicit knowledge in the classroom. It has to be an education that incorporates different forms of experiential learning. Without delving into the different theories of experiential learning, it is the form of learning which involves "the process whereby knowledge is created through the transformation of experience" (Kolb, 1984). The combination of experiential learning and entrepreneurial problem-solving as discussed above has the capacity of producing an education capable of human capital formation, an education that is not just focused on schooling that award certificates but on learning that is capable of solving societal problems and challenges.

Mass Education

Another important recommendation is for African countries to put in place policies and programs that encourage and enable mass education of their vast populations. The growing number of out-of-school children in Africa, notably in Sub-Saharan Africa (SSA), is a cause for concern. It has been estimated that over 100 million children between the ages of 6 and 18 years are out of school in SSA (Dharamshi et al., 2022; UNESCO, 2023a). This is the typical age range of the population that should be in primary and secondary schools. Given Africa's population and current high population growth rate, this cause for concern becomes even more worrisome and pertinent.

The proposition from many quarters have been that Africa needs to focus on providing access to universal primary and secondary school education. Such recommendations hammer on the current low school enrolment rates at the primary and secondary school levels of education in many African countries. As a result, the call is often made for African countries to have programs aimed at achieving universal primary and secondary school education. These types of propositions are indeed laudable as primary and secondary school education provide the foundations for higher education. Education at these two basic levels also provides students with knowledge and skills such as numeracy, basic reading and writing literacy, speaking and communication, and other foundational life skills needed for living.

However, propositions like these are largely incomplete. They will not serve the sustainable development that Africa needs. At best, such recommendations that emphasize a focus on providing universal primary and secondary school education serve as starting points to human capital development. Indeed, it is a commendable starting point for any country, especially one that is lagging behind in human capital development and other forms of development. Providing mass education will require making quality education available at the three tiers of formal education through the formal school system. However, the aim should be providing equal and universal access to public school education at all three tiers of education and making tertiary education compulsory and not just at the primary and secondary school levels.

Understandably, starting at providing universal primary and secondary school education is the most feasible, especially given resource and other constraints. However, access to tertiary education should also be provided and made compulsory for all and not the exclusive of a few. Tertiary or higher education can be provided in myriad forms and not necessarily strictly in universities or polytechnics. In the case study of Singapore, explored in Chap. 6 of this book, vocational schools were established as a means of fast-tracking human capital development to fill the skills gaps of the population. Similarly, in Finland, explored in Chap. 7, vocational schools are part of the post-basic school education. Countries such as Germany also have formal tertiary educational forms in the form of apprenticeship system. The essence is to ensure that a form of postsecondary school or post-basic school education that enables people to be economically productive is universally available.

Consequently, tertiary education in the form of vocational schools, apprenticeships, or any other form that equips students with knowledge and skills that enable economic participation through diverse occupations and professions should be provided. Each country will have to develop diverse alternate forms of tertiary education that can rapidly enhance human capital formation and development without necessarily focusing on the regular forms of higher education. These can be in form of vocational and trade schools and diverse forms of formal apprenticeships and integrated internships with domestic and foreign multinational corporations. In this light, countries can look inward toward available indigenous knowledge systems to create formal varied forms of tertiary education beyond the conventional forms transmitted to African countries through the colonial system and era. This will have the added advantages of formalizing and institutionalizing knowledge systems of trades and occupations many of which are currently largely informal, uncoordinated, unregulated, and lacking in standards in many African countries.

Adeola (2021) and colleagues explore the dynamics and intricacies of the traditional business school system of the Igbo tribe of Nigeria, the Igba-Boi system. The phenomenon they explore, a form of traditional apprenticeship system, is one that was actually quite common in some other parts of Africa before its neglect and relegation to the current conventional education form. Thankfully, for peculiar reasons, the traditional apprenticeship system was never abandoned by the Igbo tribe. Rather, it was embraced and encouraged to become more widespread and further developed into what Adeola (2021) refers to as the Igbo Traditional Business School (I-TBS). It is also one that typifies how looking inward to *some* indigenous African knowledge systems can begin to form the bedrocks of other forms of formal tertiary education that can enable human capital formation for the sustainable development of African countries.

Agriculture and manufacturing are two sectors that traditionally employ large numbers of people. In the past, and still in some places, basic primary and secondary school education was sufficient to be employed in these two industries. However, technological advancement is changing even within these two sectors. Consequently, basic education that was hitherto sufficient for effective employment within these two sectors is fast becoming insufficient.

Summarily, equal access to quality education at the three levels of education in public schools should be a right of everyone and not the privilege of a select few who can pay for it by accessing it through private schools. This of course makes two assumptions. The first is that the public schools available are in most part free and universally available and accessible. The second assumption is that education in private schools is of much higher quality than education in public schools which is often the case in certain instances in many countries (Daviet, 2023). Regardless, this is in no way advocating for the abandonment of private schools. However, quality education at all levels of education should also be accessible to all. This is one way that African countries can accelerate human capital formation and accumulation that can lead to sustainable development. Although the USA's and Europe's increase in human capital that contributed to their development was through massive increase in formal education in primary and secondary schools (Goldin, 2001, 2016), African countries need much more than that. African countries need massive increase in primary, secondary, and tertiary education. African countries cannot afford to limit universal and compulsory education to only primary and secondary school education levels, if they are to develop, and develop sustainably. Otherwise, they may continue to play catch-up to other countries in terms of development. Implicitly, countries that do not already have compulsory primary and secondary schooling policies need to enact same and move toward compulsory tertiary education. This translates to having policies that stipulate a minimum number of years of compulsory formal education at all three tiers of education.

Many African countries still do not have a comprehensive framework on compulsory education. Data from UNESCO as at 2022 shows that the average number of years of compulsory education in SSA is 8 years, while that for North Africa is 9 years. Only Egypt and Kenya have up to 12 years of compulsory formal schooling (World Bank, 2023b). Most African countries have only 7 years of compulsory schooling (Carr, 2022; World Bank, 2023b). Article 26 of the United Nations Universal Declaration of Human Rights (UDHR) states that "everyone has the right to education." This actually implies formal education or formal schooling. The article also states that "elementary education shall be compulsory." However, if African countries are to make real progress in sustainable development, they must go beyond the minimum of elementary education stated many decades ago in the UDHR. Thankfully, more recently, among the targets for United Nations (UN) Sustainable Development Goal (SDG) number 4 on quality education is to ensure that all boys and girls complete (formal) primary and secondary school education. Still, African countries need to go beyond this laudable goal to striving to ensure that a form of tertiary education is available for all. This is one area African countries cannot afford to follow the examples of other countries who have followed the path of setting compulsory schooling only at elementary education.

Achieving mass education at all three tiers of formal education will entail a rethinking of the traditional approach to formal education where a limited number of people are placed within the four walls of a classroom for learning. Nontraditional methods and approaches such as those that encourage independent, course-based, long-distance learning, apprenticeships, and digital technologies can be adapted in various forms to ensure a larger number of people are reached. A mixture of full-time and part-time compulsory schooling forms are all approaches that can be utilized in this quest for mass education. Education technology, especially those that can be deployed through phones, can also be adopted and adapted in designing forms and systems of education that aim toward enabling mass education, even if it is unconventional nontraditional classroom learning formats. The main goal is to ensure that there are means to dispense formal education that has been made compulsory to the relevant age groups. Mass education must also encompass reaching the rural areas of the continent where access to education and health care is usually a formidable challenge with poor and inadequate infrastructure, teachers, and other necessary resources for learning and care.

A Management, Science, Technology, Engineering, and Mathematics (M-STEM) Focus A lot has been written on the need for African countries to prioritize education in STEM. The need for education in STEM is further reiterated here especially because the world is at the threshold of the Fourth Industrial Revolution (4IR) and the associated digital technologies, notably artificial intelligence (AI) and associated technologies. Africa's burgeoning youthful population cannot afford to be left behind in the new dispensation. Moreover, 4IR also portends that the KSAOs that can and will be

regarded as human capital are bound to change in the nearest future. A mass education that will truly result in human capital must be one that focuses on STEM and ensures that the curriculum can deliver digital literacy. In addition to STEM, management is a core area that African countries need because education in STEM without the requisite management education to deliver leadership, management, and organization skills will be underproductive.

Regular Predictive Curriculum Update The content of what and how students are learning needs to also be regularly updated to keep up with the technological times—recognizing the technological advancements of the times. It is not enough simply to provide education that enables numeracy and reading literacy. The need for reading and numerical literacy has given way to the additional necessity of digital literacy. Times have changed and is continually changing rather rapidly. At a time of fast-paced innovations in science and (digital) technology that is leading to what Peters (2017) aptly refers to as "an age of automated cognition," the illiterate of today is not one that cannot read and write but one that cannot make adequate use of modern-day digital technologies to communicate and live productive lives. Consequently, efforts must be made to periodically update curriculum toward ensuring that it is able to deliver learning that is relevant both now and in the predictable future. This can be incorporated into and combined with the regular human capital audit discussed above.

Quality Teachers, Teaching, and the Teaching Profession

UN's SDG 4 is on quality education and aims to "ensure inclusive and equitable quality education and promote lifelong learning opportunities for all." The issue of quality education is complex. Several factors such as number of students and classroom size; textbooks and teaching materials;

teachers' knowledge content, training, and management; and physical, ICT, and other infrastructure all contribute to quality education (Global Education Monitoring, 2022; Olaniyan, 2022). However, having quality teachers and teaching are fundamental foundational components of quality education. As the case study on Finland in Chap. 7 highlights, quality teachers are needed to ensure quality education, and this depends largely on quality teachers' education and having a high regard for the teaching profession such that people are attracted to the teaching profession. Having teachers that willingly join the teaching profession and equipping them with the capacity to teach effectively, as clearly shown in the case of teachers in Finland, are twin factors to having quality teachers. Teachers should be required to have a minimum level of tertiary education even to teach in primary schools, at least a bachelor's degree.

Unfortunately, across Africa, the teaching profession is one that is not highly regarded and is coupled with poor remuneration and welfare conditions (Evans & Yuan, 2018; Miiro, 2022; See et al., 2022). Consequently, many people are reluctant to become teachers particularly at the primary and secondary school levels. Even at the tertiary level, joining the teaching profession is generally regarded as a last resort because of the welfare and working conditions. Consequently, getting people that are willing, not compelled, to become teachers and then enabling them with the capacity to teach can be herculean tasks. According to a UNESCO (2023b) report, over 15 million teachers are still needed in Africa. Clearly, attracting people to the teaching profession to teach in schools, particularly primary and secondary schools in Africa, is a crucial need that requires urgent action. This will definitely entail improving the welfare and working conditions of teachers. More than these, improving the status of the teaching profession is a necessity. The huge concern is how this can be done amid funding and other resource constraints in many African countries. However, beyond welfare and working conditions that require funding, a starting point is improving the status of the teaching profession.

The Teaching Profession Making the teaching profession attractive can begin by professionalizing it. Although reference is often made to teaching as a profession, in many African countries and indeed in many countries globally, teaching is seldom seen as truly a "profession." There have been debates, originating from decades ago and still ongoing, about whether teaching is truly a profession or an occupation or even just a job or perhaps even a "calling" (Marshall, 1957; Merrow, 2021; Runte, 1995). Without delving into this debate, Finland has inadvertently provided empirical evidence of the positive possibilities that can result from professionalizing the teaching profession. Reforms in the Finnish education system may not have had the explicit goal of making teaching a profession. However, what the reforms achieved by making teaching more attractive can be likened to a form of professionalization. Putting in place regulatory frameworks that consist of rigorous qualifications for teaching in primary and secondary schools similar to the requirements for lecturing, or teaching, in universities is a necessary step toward professionalization of teaching. Teaching needs to become a more valued profession. Ideally, only the most "talented" individuals in a country should become teachers. The case of Finland has shown that when the most "talented" are the crux of the teaching profession, the learning outcomes are better. In addition to minimum normal formal education requirements, teachers should be required to have professional certifications administered by regulated professional bodies. This will go a long way in professionalizing teaching and improving the status of teachers in the society because it debunks the notion that just anyone can become a teacher.

Many arguments have been made for teachers' continuous professional development to enhance teachers' knowledge, quality, and teaching. In addition to these, continuous professional development should be a general requirement not only because it directly impacts teaching quality but also because it has the inadvertent benefit of improving the status of the teaching profession and contributing to its professionalization.

Workplace Training and Development

Human capital development goes beyond the formal school education system. As discussed in the previous chapter, the formal education system is only one aspect of the internal development approach of accumulation human capital, albeit a major and nonnegotiable one. Countries that are serious about human capital development also have policies and programs that go beyond the formal education system. The case studies of Singapore, Finland, and Canada, presented in this book, have great formal education systems. These countries also have a holistic educational system that extend beyond the formal education system. The educational system is somewhat integrated toward ensuring that people continue to acquire human capital beyond what has been acquired through the formal education system. The aim is to ensure that after formal schooling in the formal education system, as people embark on working in the workplace, they also continue to acquire necessary KSAOs in addition to other specific human capital acquired that may be acquired on the job. Organizations and workplaces should be encouraged to provide training and development to their workers. This can be achieved through giving some forms of tax credits and other inducements that incentivize organizations to send their employees periodically for training.

Health-Care Systems for All

As a vital component of human capital, health is sometimes overlooked in the discussion on human capital development. In fact, some global human capital indexes[3] do not include any indicators for measuring health. Without good health, people cannot put their knowledge, skills, and any other ability they have to productive use. Hence, human capital is incomplete without good health, and a country's human capital development system cannot be complete without considering the health-care system. Ideally, a country's health-care system should be one that is in tandem with its formal education system such that both systems form a coherent whole. However, in reality, countries scarcely design their education systems and health-care systems jointly. Rather, both evolve separately over time, and attention is rarely given to how one affects the other.

Recently, the main recommendation, especially from global agencies, has been for countries to have a form of universal health-care system that guarantees universal health coverage, one where everyone has access to the "quality health services they need, when and where they need them without financial hardship (WHO, n.d.)." Historically, advocates of universal health-care coverage have moved from recommending universal primary health care to a universal health-care system that covers all forms of health-care services, from primary health care to tertiary health-care services (Ranabhat et al., 2023). This is largely due from sustainable development goal 3 (SDG 3). Universal health coverage is stated as the eighth target under SDG 3. A core element of a universal health-care system is that people do not have to pay out of pockets for health-care services. The essence of a universal health-care system is not the provision of quality health care alone. Neither is it the mere provision of public health-care services. A universal health-care system is to ensure that people do not need to worry about payment at the point of receiving health care. Target 3.8 of SDG 3 emphasizes essential health-care services, medicines, and vaccines, while other targets of SDG 3 speak to other health issues. Consequently, universal health-care coverage can range from providing basic primary health care to a whole range of other health-care services as advocated by some global agencies. There are also different models of universal health-care systems in theory and in practice. Worldwide, countries with universal health-care systems such as Australia, Finland, Canada, and the United Kingdom (UK) all have different models of universal health coverage.

---

[3] See Chap. 5 for a comprehensive discussion on available global human capital indexes.

Given the different models of universal health-care systems, what form of a universal health-care system should African countries adopt?

A good universal health-care system is one that works and guarantees access to a minimum level of health-care services. It is one that is capable of meeting the health needs of the population of the country given each country's peculiar context. African countries should have universal health-care systems that at the very minimum guarantee the provision of quality primary and essential health services for all. As expressed by Martin et al. (2018), universal health coverage is not a destination but an aspiration, and countries must continuously consider the depth and scope of coverage that is achievable and fiscally feasible. The UK's form of universal health-care system, the National Health Service (NHS), or Canada's health-care system, Medicare, or that of any country for that matter, may not be ideal for particular African countries. Consequently, while African countries should have universal health-care systems that guarantee universal health coverage at certain health-care levels, the right path does not lie in adopting or even adapting another country's health-care system. Universal health-care systems usually evolve over time. It is not feasible for a country that currently has no holistic health-care system in place to jump to having a universal health-care system that covers all ranges of health-care services. Rather, countries should aim at starting from their current health-care system to a universal health-care system that provides essential primary health services.

Human Capital Migration and Immigration

Immigration remains one of the viable ways of accessing and accumulating human capital open to countries because it gives access to people with human capital, subsequently helping in human capital acquisition and accumulation. It is no coincidence that many developed countries continue to design special immigration programs to attract skilled and highly skilled people to their countries. There is a stream of research that has established the positive impact of immigration of highly skilled people on the economic growth of destination countries (e.g. Barro & Lee, 2013; Ehrlich & Kim, 2015; Di Maria & Stryszowski, 2009). Despite ongoing debates in different research streams on the impact of such immigration of highly skilled people on origin countries, the fact is that immigration actually helps in human capital formation in destination countries—there is no debate on this in these extant research streams. Consequently, the positive impact of immigration of highly skilled people on economic growth of destination countries that have been established by various studies is not surprising. Indeed, it is expected. The surprise would have been finding the opposite impact or no impact in research findings.

However, using immigration as a means of accessing people with human capital and in human capital accumulation requires a careful strategy and proper implementation to be successful. The plan must be designed to attract people with human capital in the areas where the country lacks expertise, and a human capital audit and continuous tracking, as discussed earlier in this chapter, can aid in understanding the skills gaps in a country and the particular areas to target. Using strategic immigration to attract human capital should be an adjunct plan to developing human capital internally. For African countries, an additional necessary recommendation is that any strategic immigration program must also have supporting policies that facilitate knowledge transfer. African countries cannot afford to use this route without ensuring that there are processes in place to facilitate managerial and technical knowledge transfer. Relatedly, countries can also access human capital through foreign direct investment (FDI). Many African countries have been on a quest to attract FDI. Despite Africa's rather low share of global FDI inflows, inflows into Africa have been rising over the years reaching a record $US83 billion in 2021 (Morgan et al., 2022; UNCTAD, 2022). The discuss on FDI inflows seldom focus on its impact on access to human capital and its formation within recipient countries. Movement of FDI between countries implicitly involves movement of human capital. Countries can access human capital and stimulate human capital formation through FDI as it can help countries to have access to people with both technical and managerial skills, two very important forms of human capital. As African countries continue to seek FDI inflows, it is pertinent to take cognizance of the types of potential human capital that are likely to accompany such financial capital inflows.

A Multi-stakeholder Approach to Human Capital Development

The United Nations Educational Scientific and Cultural Organizations (UNESCO) recommends that member countries should allocate 4–6% of their gross domestic product (GDP) or 15–20% of their budgets to education (UNESCO, 2021). Many African countries are yet to comply with this recommendation. Available data from UNESCO for 2021, for example, show that 19 African countries spent at least 4% of their GDP on education, while 24 African countries spent at least 15% of their public expenditure on education.[4] Fourteen countries belong to both categories, a signal of the value they place on education as a means of developing human capital. Many developed countries surpass these recommended threshold spending on education. Considering that the GDP of many of the more developed countries already surpassed that of many African countries in absolute amounts, the monies earmarked on education by African countries, in comparison to these countries, is meager. Clearly,

remedying the poor state of human capital in Africa requires increased funding, and from available data, African countries need to spend more on education and of course on health care.

Given funding constraints, it is quite difficult to expect the government to be the sole spender on education and human capital development. The government is expected to design policy frameworks and provide adequate regulations that embed an overall strategy on human capital development. Funding human capital development and inputs into the requisite policy frameworks require multiple stakeholders apart from governments. Organizations in the private and third sectors and the different communities of Africans in diaspora can all play active roles in human capital development. Governments need to deliberately increase the involvement of other stakeholders other than global agencies in developing their countries' human capital. Global agencies appear to emphasize an increase in spending on education and health as solution to the challenges of human capital development in Africa. To show its commitment to Africa's human capital development, the World Bank launched the Africa Human Capital Plan in 2019. In 2021, just after the COVID-19 global pandemic, the World Bank reportedly spent over $US8.96 billion to finance human capital development in African countries, and for the 3 years leading up to 2022, a total of $US34.3 billion dollars was spent on different projects on the continent (World Bank, 2021, 2022). This is commendable, and the monies spent signal a great commitment to human capital development in African countries. Nevertheless, the challenge of human capital development in Africa cannot simply be solved by devoting vast amounts of money to it. Improving human capital in African countries goes beyond increased funding. There are other pertinent factors to consider apart from funding such as expenditure efficiency, accountability, and the kind of projects monies are spent on. Public expenditure on education in Africa has been riddled with inefficiencies as there have been reports of the continent having the worst education spending efficiency globally (AfDB, 2020; Sikayena et al., 2022). Huge spending on

---

[4]Countries that spent at least 4% of their GDP on education: Namibia, Sierra Leone, Lesotho, Mozambique, South Africa, Algeria, Cape Verde, Seychelles, Morocco, Rwanda, Senegal, Eswatini, Burkina Faso, Burundi, Mauritius, Kenya, Mali, Togo, and Ethiopia. Countries that spent at least 15% of their public expenditure on education: Sierra Leone, Namibia, Morocco, Ethiopia, Democratic Republic of Congo, Burkina Faso, Senegal, Burundi, Mozambique, South Africa, Congo Republic, Benin, Togo, Cote d'Ivoire, Mali, Algeria, Chad, Gambia, Rwanda, Madagascar, Gabo, Eswatini, Cape Verde, and Cameroun; See Chap. 9 for graphical and comparative analyses of government expenditure on education.

human capital projects may not be sustainable or have the necessary impact if the right national and sub-national policies and programs are not in place. Moreover, when it comes to human capital, just like other matters, the focus and priority of global agencies can be quite different from that of national and sub-national governments. This can result in a misalignment in purpose and objectives and impede overall progress. None should however preclude the other as multiple stakeholders are needed for substantial progress to be made in developing Africa's human capital. Nongovernmental stakeholders usually demand more accountability and transparency than government-related stakeholders. Consequently, involving other stakeholders in human capital development efforts apart from governments alone may increase efficiency.

A viable option is for human capital development to be funded and supported by multiple stakeholders, particularly through strategic partnerships. While this is already going on, more is needed, and current efforts seem to be largely fragmented. A coordinated effort is required in each country. However, it can be difficult to coordinate efforts without national and local government strategic plans on human capital development. Still, increasing the funding of education and health care will not automatically translate into better human capital development without the necessary framework for accountability and transparency. Civil society organizations that monitor governments and private sector investments in education and health are direly needed. Africans in the diaspora, in additional to making financial investments in human capital projects on the continent, can play a huge part in this regard.

Conclusion: Summary of Recommendations

As time evolves, human capital evolves because the knowledge, skills, abilities, and other attributes that have economic value change. Therefore, for countries whose human capital stock levels are still low, such as African coun-

tries, increasing the rate of human capital accumulation requires a deliberate action plan that is implemented, monitored, and updated regularly. Below is a summary list of the core recommendations for African countries that have been discussed in this chapter.

I. Have a strategic human capital development plan that will serve as a guide to policies and programs on human capital development.

II. Conduct a human capital audit to take stock of the human capital available to the country, and continue tracking the rate of progress of human capital accumulation through continuous human capital audits.

III. Provide mass education at all levels of education, and focus on education in M-STEM and digital knowledge and skills.

IV. Ensure the education system promotes entrepreneurship and incorporates entrepreneurial problem-solving.

V. Conduct regular curriculum updates at all three tiers of education.

VI. Put in place national policy frameworks to incentivize workplace training and development.

VII. Professionalize the teaching profession.

VIII. Involve multiple stakeholders in human capital development.

IX. Work toward establishing universal healthcare systems that provide essential health care at the minimum.

X. Have strategic immigration programs to attract people with required human capital if necessary, and ensure that there is an accompanying knowledge transfer framework.

While not exhaustive, implementing these recommendations will place countries on a path to accumulating human capital and toward attaining long-term and sustainable development. Given economic and political stability, African countries can adopt or adapt these recommendations for sustainable and long-term development. Human capital development and accumulation is a long-term and a "never-ending" project which is why developed countries continue to make the necessary investments in the human capital of today and potential human capital of the future.

References

Adams, S. (2009). Foreign direct investment, domestic investment, and economic growth in Sub-Saharan Africa. *Journal of Policy Modelling, 31*(6), 939–914. https://doi.org/10.1016/j.jpolmod.2009.03.003

Adeola, O. (2021). *Indigenous African enterprise: The Igbo Traditional Business School (I-TBS)* (Advanced series in management) (Vol. 26). Emerald Publishing.

AfDB. (2020). *African economic outlook 2020: Developing Africa's workforce for the future.* Report of the African Development Bank Group. https://www.afdb.org/en/documents/african-economic-outlook-2020

Alfaro, L., Chanda, A., Kalemli-Ozcan, S., & Sayek, S. (2000). FDI and economic growth: The role of local financial markets. *Journal of International Economics, 64*(1), 89–112. https://doi.org/10.1016/S0022-1996(03)00081-3

Anetor, F. O., Esho, E., & Verhoef, G. (2020). The impact of foreign direct investment, foreign aid and trade on poverty reduction: Evidence from Sub-Saharan African countries. *Cogent Economics & Finance, 8*(1). https://doi.org/10.1080/23322039.2020.173734

Archer, E., & Chetty, Y. (2013). Graduate employability: Conceptualization and findings from the University of South Africa. *Progression, 35*(1), 136–167.

Aziegbe-Esho, E., & Anetor, F. O. (2020). Religious organisations and quality education for African women: The case of Nigeria. In O. Adeola (Ed.), *Empowering African women for sustainable development: Towards achieving the United Nations' 2030 goals* (pp. 73–83). Palgrave Macmillan.

Barro, R. J., & Lee, J. W. (2013). A new data set of educational attainment in the world, 1950–2010. *Journal of Development Economics, 104*, 184–198. https://doi.org/10.1016/j.jdeveco.2012.10.001

Bénétrix, A., Pallan, H., & Panizza, U. (2023). *The elusive link between FDI and economic growth.* World Bank Policy Research Working Paper 10422. https://doi.org/10.1596/1813-9450-10422

Bruno, R. L., Campos, N. F., & Estrin, S. (2018). Taking stock of firm-level and country-level benefits from foreign direct investments. *Multinational Business Review, 26*(2), 126–144. https://doi.org/10.1108/MBR-02-2018-0011

Carr, O. G. (2022). Promoting priorities: Explaining the adoption of compulsory schooling laws in Africa. *International Journal of Educational Development, 88*, 102523. https://doi.org/10.1016/j.ijedudev.2021.102523

Carter, E. W., Awsumb, J. M., Schutz, M. A., & McMillan, E. D. (2021). Preparing youth for the world of work: Educator perspectives on pre-employment transition services. *Career Development and Transition for Exceptional Individuals, 44*(3), 161–173.

Casillas, A., Kyllonen, P., & Way, J. (2019). Preparing students for the future of work: A formative assessment approach. In F. L. Oswald, T. S. Behrend, & L. L. Foster (Eds.), *Workforce readiness and the future of work* (pp. 35–52). Taylor, & Francis/Routledge.

Chamorro-Premuzic, T., & Frankiewicz, B. (2019). Does higher education still prepare people for jobs? *Harvard Business Review.* https://hbr.org/2019/01/does-higher-education-still-prepare-people-for-jobs

Cheng, M., Adekola, O., Albia, J., & Cai, S. (2022). Employability in higher education: A review of key stakeholders' perspectives. *Higher Education and Development, 16*(1), 16–31.

Daviet, B. (2023). Do private schools provide a better quality of education? Beyond the private – Public divide. A literature review 1981–2022. *Cahiers de la Recherche sur Ll'éducation et les Savoirs, 22*, 33–58.

Dharamshi, A., Barakat, B., Antoninis, M., Montoya, S., & UNESCO. (2022). *A bayesian cohort model for estimating SDG indicator 4.1.4: Out-of school rates.* https://www.unesco.org/gem-report/sites/default/files/medias/fichiers/2022/08/OOS_Proposal.pdf

Di Maria, C., & Stryszowski, P. (2009). Migration, human capital accumulation and economic development. *Journal of Development Economics, 90*(2), 306–313. https://doi.org/10.1016/j.jdeveco.2008.06.008

Ehrlich, I., & Kim, J. (2015). Immigration, human capital formation, and endogenous economic growth. *Journal of Human Capital, 9*(4), 518–563. https://www.jstor.org/stable/26456399

Evans, D. K., & Yuan, F. (2018). *The working conditions of teachers in low- and middle-income countries.* Background Paper, World Bank's 2018 World Development Report. "Realizing the promise of education for development".

Global Education Monitoring. (2022). *BRIEFLY: Measuring education quality in Africa.* Global Education Monitoring (GEM) Report. Available at https://world-education-blog.org/2012/05/07/briefly-measuring-education-quality-in-africa/

Goldin, C. (2001). The human capital century and American leadership: Virtues of the past. *Journal of Economic History, 61*, 263–291.

Goldin, C. (2016). Human capital. In C. Diebolt & M. Haupert (Eds.), *Handbook of cliometrics.* Springer.

Goldin, C., & Katz, L. F. (2008). *The race between education and technology.* The Belknap Press of Harvard University.

Grosemans, I., De Cuyper, N., Forrier, A., & Vansteenkiste, S. (2023). Graduation is not the end, it is just the beginning: Change in perceived employability in the transition associated with graduation. *Journal of Vocational Behavior, 145*, 103915. https://doi.org/10.1016/j.jvb.2023.103915

Hanushek, E. A., & Woessmann, L. (2008). The role of cognitive skills in economic development. *Journal of Economic Literature, 46*(3), 607–668.

Igbokwe-Ibeto, C. J., Agbodike, F. C., & Osakede, K. O. (2018). Entrepreneurial curriculum in African universities: A panacea to graduates' unemployment if? *Africa's Public Service Delivery and Performance Review, 6*(1), a222. https://doi.org/10.4102/apsdpr.v6i1.222

King, E. M., & Winthrop, R. (2015). *Today's challenges for girls' education* (Working Paper 90). Brookings Global Economy and Development.

Kolb, D. A. (1984). *Experiential learning: Experience as the source of learning and development.* Prentice-Hall.

Marshall, M. S. (1957). Is teaching a profession? *Improving College and University Teaching, 5*(4), 88–92. https://doi.org/10.1080/00193089.1957.10533980

Martin, D., Miller, A. P., Quesnel-Vallée, A., Caron, N. C., Vissandjée, B., & Marchildon, G. P. (2018). Canada's health-care system: Achieving its potential. *Lancet, 391*(10131), 1718–1735. https://doi.org/10.1016/S0140-6736(18)30181-8

Merrow, J. (2021). *Is teaching a profession, an occupation, a calling, or a job?* The Merrow Report. Available at https://themerrowreport.com/2021/05/28/is-teaching-a-profession-an-occupation-a-calling-or-a-job-2/

Mgaiwa, S. J. (2021). Fostering graduate employability: Rethinking Tanzania's University practices. *SAGE Open, 11*(2). https://doi.org/10.1177/21582440211006709

Miiro, F. (2022). The teaching profession in Africa: Challenges and prospects. *Interdisciplinary Journal of Education, 5*(2), 117–124. https://doi.org/10.53449/ije.v5i2.208

Morgan, S., Farris, J., & Johnson, M. E. (2022). *Foreign direct investment in Africa: Recent trends leading up to the African Continental Free Trade Agreement (AfCFTA).* Economic Research Service, Economic Information Bulletin Number 242. https://www.ers.usda.gov/webdocs/publications/104996/eib-242.pdf?v=3027.7

Njuguna, A. E., & Nnadozie, E. (2022). Investment climate and foreign direct investment in Africa: The role of ease of doing business. *Journal of African Trade, 9*, 23–46. https://doi.org/10.1007/s44232-022-00003-x

Obor, O. D., & Kayode, D. I. (2022). Highly educated but unemployable. In R. Baikady, S. Sajid, J. Przeperski, V. Nadesan, I. Rezaul, & J. Gao (Eds.), *The Palgrave handbook of global social problems.* Palgrave Macmillan. https://doi.org/10.1007/978-3-030-68127-2_162-1

Olaniyan, O. (2022). *Improving the quality of teachers and teaching across African countries.* United Nations. https://www.un.org/development/desa/pd/sites/www.un.org.development.desa.pd/files/undesa_pd_2022_egm3_presentation_olaniyan.pdf

Olutuase, S. O., Brijlal, P., Yan, B., & Ologundudu, E. (2018). Entrepreneurial orientation and intention: Impact of entrepreneurial ecosystem factors. *Journal of Entrepreneurship Education, 21*(Special Issue), 1–14.

Olutuase, S. O., Brijlal, P., & Yan, B. (2023). Model for stimulating entrepreneurial skills through entrepreneurship education in an African context. *Journal of Small Business & Entrepreneurship, 35*(2), 263–283. https://doi.org/10.1080/08276331.2020.1786645

Panth, B., & Maclean, R. (2020). Introductory overview: Anticipating and preparing for emerging skills and jobs—Issues, concerns, and prospects. In B. Panth & R. Maclean (Eds.), *Anticipating and preparing for emerging skills and jobs. Education in the Asia-Pacific region: Issues, concerns and prospects* (Vol. 55). Springer. https://doi.org/10.1007/978-981-15-7018-6_1

Peters, M. A. (2017). Technological unemployment: Educating for the fourth industrial revolution. *Educational Philosophy and Theory, 49*(1), 1–6. https://doi.org/10.1080/00131857.2016.1177412

Ranabhat, C. L., Acharya, S. P., Adhikari, C., & Kim, C. (2023). Universal health coverage evolution, ongoing trend, and future challenge: A conceptual and historical policy review. *Frontiers in Public Health, 11*, 1041459.

Runte, R. (1995). Is teaching a profession? In G. Taylor & R. Runte (Eds.), *Thinking about teaching: An introduction.* Harcourt Brace Publishers.

See, B. H., Munthe, E., Ross, S. A., Hitt, L., & El Soufi, N. (2022). Who becomes a teacher and why? *Review of Education, 10*(3), e3377.

Siivonen, P., Isopahkala-Bouret, U., Tomlinson, M., Korhonen, M., & Haltia, N. (2023). Introduction: Rethinking graduate employability in context. In P. Siivonen, U. Isopahkala-Bouret, M. Tomlinson, M. Korhonen, & N. Haltia (Eds.), *Rethinking graduate employability in context.* Palgrave Macmillan. https://doi.org/10.1007/978-3-031-20653-5_1

Sikayena, I., Bentum-Ennin, I., Andoh, F. K., & Asravor, R. (2022). Efficiency of public spending on human capital in Africa. *Cogent Economics & Finance, 10*, 1. https://doi.org/10.1080/23322039.2022.2140905

Stanley, A. (2023). *African century.* IMF: International Monetary Fund, United States of America. Available at https://policycommons.net/artifacts/4809842/african-century/5646416/ on 26 Jan 2024. CID: 20.500.12592/n4xg81.

Su, Y., & Liu, Z. (2016). The impact of foreign direct investment and human capital on economic growth: Evidence from Chinese cities. *China Economic Review, 37*, 97–109. https://doi.org/10.1016/j.chieco.2015.12.007

UN. (2022). *World population prospects.* Available at https://population.un.org/wpp/Download/Standard/Population/

UNCTAD. (2022). *Investment flows to Africa reached a record $83 billion in 2021.* United Nations Conference on Trade and Development (UNCTAD). Available at https://unctad.org/news/investment-flows-africa-reached-record-83-billion-2021

UNESCO. (2021). *UNESCO member states unite to increase investment in education.* Available at https://www.unesco.org/en/articles/unesco-member-states-unite-increase-investment-education

UNESCO. (2023a). *Out-of-school rate.* Available at https://education-estimates.org/out-of-school/

UNESCO. (2023b). *The persistent teacher gap in Sub-Saharan Africa is jeopardizing education recovery.* Available at https://www.unesco.org/en/articles/persistent-teacher-gap-sub-saharan-africa-jeopardizing-education-recovery

Uzoka, F. A. (2006). Challenges in entrepreneurship in home economics education. *Nigeria Journal of Education, 4*(2), 98–106.

Verhoef, G. (2017). *The history of business in Africa: Complex discontinuity to emerging markets.* Springer International.

WHO. (n.d.). *Universal Health Coverage (UHC).* Available at https://www.who.int/news-room/fact-sheets/detail/universal-health-coverage-(uhc)

World Bank. (2021). *Human capital project: Year 3 progress report.*

World Bank. (2022). *Putting people first: The Africa human capital year 3 progress report.* Available at https://ahcp.worldbank.org/en/

World Bank. (2023a). *Investing in youth, transforming Africa.* Available at https://www.worldbank.org/en/news/feature/2023/06/27/investing-in-youth-transforming-afe-africa

World Bank. (2023b). *Compulsory education, duration (years), -Sub-Saharan Africa.* UNESCO Institute for Statistics (UIS). UIS.Stat Bulk Data Download Service. apiportal.uis.unesco.org/bdds. Accessed 19 Sept 2023.

A Strategic Human Capital Approach to the Sustainable Development of African Countries: Conclusion

12

Abstract

A strategic human capital approach to Africa's sustainable development is anchored on prioritizing investment in people's education and health above all other investments and ensuring that other investments are threaded around their impact on people, their human capital development, and how they facilitate people's human capital deployment in economic productive activities. This chapter reiterates the merits of such an approach to sustainable development and why it should be the main developmental path for African countries. This concluding chapter of the book also identifies research opportunities in different areas that could provide further insights into the sustainable development of African countries.

Keywords

Sustainable development · SDGs · Agenda 2063 · Human capital · African countries

Introduction

There is often much talk and discuss about the vast natural resources of the African continent. Reference is readily made to minerals beneath the ground and the enormous potential for agriculture, given the millions of hectares of uncultivated land that is still on the continent. Attention is seldom paid to another form of resource that is abundant and increasing in Africa—the continent's large youthful and growing population. The outcome of having a huge youthful population, one that has been predicted to continue to increase in the forthcoming years, depends on the investments in people that are made today. This chapter summarizes the author's ideas on adopting a strategic human capital approach to development presented in this book. Table 12.1 summarizes the main ideas and contents of each chapter. The chapter also presents some of the hindrances that are capable of limiting the implementation of the recommendations provided in previous chapters and charts a roadmap for further research and scholarship on human capital and Africa's sustainable development.

The idea reiterated throughout the chapters of this book is that taking a strategic human capital approach is the viable path toward sustainable development for African countries. This approach to Africa's sustainable development is anchored on prioritizing investment in people's education and health above all other investments and ensuring that other investments are threaded around their impact on people, people's human capital development, and deployment. In this approach, the building of physical infrastructure, for example, is prioritized based on whether they can

Table 12.1 Summary of chapters of the book

Part	Chapter and chapter title	Summary/main idea(s)/lessons
	1 A Strategic Human Capital Approach to the Sustainable Development of African Countries: Introduction	This chapter introduces the book and the strategic human capital approach to sustainable development which is the deliberate and coordinated investment in the education and health of the people and the assessment of other investments by their relations and impact on the productive capacities and well-being of the people.
1 Human Capital: Laying the Foundation	2 What Is Human Capital?	This chapter defines and explains the meaning of human capital. Human capital is the knowledge, skills, abilities, and other characteristics (KSAOs) of individuals that can be put to productive use and have economic value.
	3 The Nature of Human Capital	This chapter examines the major components, sources, types, and levels of human capital using a knowledge-based conceptual framework. Human capital can be at the individual level and at aggregate levels such as organizations, societies, and countries. Aggregate levels of specific types of human capital can emerge such that countries can have national human capital.
	4 Outcomes and Benefits of Human Capital	This chapter discusses the various benefits of human capital to individual persons and groups. For individuals, human capital increases wages and earnings and job performance and improves the quality of life, health, and Well-being and the capacity to innovate and be successful at entrepreneurship. At more aggregate levels, human capital improves organizational performance, contributes substantially to economic growth and country's competitiveness, and facilitates gains to trade and foreign direct investment (FDI).
2 Case Studies of National Outcomes of Strategic Human Capital Development	5 Global Human Capital Indexes and Introduction to Part 2	This chapter presents and explains different global human capital indexes and the indicators used in measuring countries' human capital in the indexes. Despite many criticisms on the global indexes for human capital, it is a good reflection of the state of a country's human capital. Countries will do well to use the indexes as a gauge and guide to their human capital status in terms of human capital development.
	6 The Emergence of Asia: The Case of Singapore	This chapter presents the first case study of a country's strategic use of human capital—Singapore. Guided by the visionary leadership of lee Kuan yew, national policies, programs, and the attraction and use of FDI were centered around human capital development: An understanding of the people and their current and potential human capital.
	7 Integrated Europe: The Case of Finland	This chapter presents the case study of Finland, a top-ranking country on almost all global human capital indexes and with a world-acclaimed formal education system. Compulsory schooling reforms that involve multi-stakeholder engagement, quality teachers' education, and the high requirements and status of the teaching profession have all contributed to the quality education system that the country has become globally renowned for.
	8 Immigration and Human Capital Accumulation in North America: The Case of Canada	Canada's historical and contemporary distinctive approach to human capital accumulation using strategic immigration and the lessons from it are presented in this chapter.

(continued)

Table 12.1 (continued)

Part	Chapter and chapter title	Summary/main idea(s)/lessons
3 Human Capital in African Countries	9 State of Africa's Human Capital and Introduction to Part 3	Data and statistics reflect the current poor state of human capital in Africa and the inherent opportunities for human capital development and accumulation. Africa's huge and growing youthful population can only yield potential demographic dividends through human capital development and accumulation.
	10 Generic Strategic Approaches to Human Capital Development and Accumulation	Using human capital theory, resource-based theory (RBT), and transaction cost economies, this chapter presents three generic approaches (internal human capital development, external human capital accumulation, and a hybrid approach) that can be adopted by African countries to develop and accumulate human capital. Internal human capital development consists of the formal education system, health-care system, informal education system, workplace training and development, and civil organizational behavior knowledge systems. External human capital accumulation consists mainly of strategic immigration and diaspora development.
	11 Developing Africa's Human Capital: Recommendations and Suggestions	African countries need to prioritize human capital development by having a national human capital development plan that incorporates conducting human capital audits and tracking. Other recommendations include having an education system that truly creates human capital through entrepreneurial problem-solving; mass education through compulsory schooling at primary, secondary, and any form of tertiary education; quality teaching through professionalizing the teaching profession; and having a form of universal health care for essential health-care services. Adopting a multi-stakeholder approach and looking inward toward alternative tertiary education forms based on and derived from some indigenous knowledge systems would help African countries in developing their human capital.
	12 A Strategic Human Capital Approach to the Sustainable Development of African Countries: Conclusion	This chapter concludes the book, reiterating some of the ideas presented, and provides some broad research opportunities.

effectively facilitate human capital development and their deployment to productive economic activities. African countries need to pay greater policy attention to human capital accumulation, particularly through human capital development in order to increase the stock and quality of human capital for today and the future.

The merits of a strategic human capital approach to development are numerous. The approach has enormous support from extant research from diverse disciplines that study human capital, though their study of the concept is often from distal points and perspectives.

Nevertheless, the evidence is conclusive—human capital has several benefits to individuals, organizations, and the society. The very idea of human capital itself resulted from the observed benefits of schooling and education and other investments in the human person to individuals and societies (Becker, 1962, 1964; Mincer, 1958; Schultz, 1961; Smith, [1776] 2009). For individuals, human capital leads to increased earnings and job performance, entrepreneurial success, and improvement in the quality of life and well-being (Becker, 2007; Bloom & Canning, 2003; Dimov, 2017; Filmer et al., 2021; Martin et al., 2013;

Unger et al., 2011). Human capital increases the performance of organizations in diverse ways (Crook et al., 2011; Hessels et al., 2020). For societies and countries, human capital affects economic growth and development and is related to new technology and technological inventions (Acemoglu & Autor, 2012; Barro, 1991; Dao & Khuc, 2023; Goldin, 2016; Goldin & Katz, 2008; Wilson & Briscoe, 2004). Indeed, diverse literature streams study human capital and its effects on individuals, organizations, societies, and countries. It is, thus, impossible to list the numerous studies.

In this post-industrial revolution modern era of economies, the economic growth of countries with greater stock of human capital outpaces that of other countries; human capital accounts for much of the difference in the economic growth and development between the USA and European countries in the early twentieth century (Goldin, 2001, 2016). Undoubtedly, human capital contributes to the growth of economies especially when it is applied to tasks in different occupations (Acemoglu & Autor, 2012). Human capital contributes to the competitiveness of countries (Porter, 1990, 2000) and facilitates the positive impacts of financial and physical capital on economic growth. The evidence is also clear of countries such as Singapore that have changed their developmental status by focusing on human capital development, albeit it may not have been explicitly stated that they were undertaking the strategic human capital approach to development. Consequently, the strategic human capital approach to Africa's sustainable development advocated in this book is not new. Rather, it has been in existence and enjoys massive support from research. What is new is the cohesive and holistic formalization of the approach. This book serves as a means of bringing the diverse strings of the merits of human capital into a united whole and puts forward an approach that can be utilized by African countries, and indeed other countries, with the desire for sustainable development.

The Strategic Human Capital Approach to Sustainable Development

This approach emphasizes a focus on increasing the stock of human capital in a country. It rests on the understanding that the true wealth of a country are the people and not the natural or mineral resources. The strategic human capital approach to development is the deliberate and coordinated investment in the education and health of the people and the assessment of other investments by their relations and impact on the productive capacities and well-being of the people. Basically, under this approach, other investments are made based on their potential impact on human capital development and deployment. It consists of accumulating and developing human capital. While both terms, human capital accumulation and human capital development, are similar, they are nonetheless quite distinct. Human capital accumulation for a country is increasing the stock of human capital by having more people with human capital. Human capital development is making investments in people to increase their human capital. Consequently, human capital accumulation can be done through human capital development or by just ensuring that the country has access to people with human capital which can be through immigration and foreign direct investments, with the associated knowledge transfer.[1] Whichever route is chosen, the essential goal is to increase the stock of human capital in the country and to harness it in combination with other resources toward reaping the benefits it brings at all levels—to individuals, organizations, and the country at large.

A strategic human capital approach to development is much more sustainable than other approaches because it puts the people at the core of development and has multiple "snowballing" multiplier effects on every sector of the economy. It is also capable of alleviating poverty. One of World Bank's interest in promoting human capital development stems from the numerous possi-

---

[1] See Chap. 10 for details of the generic approaches to human capital accumulation and development.

ble benefits. In fact, this approach is implicitly the mainstay of development in the modern economy, particularly in the increasingly knowledge-based and digital economy. Much progress can be made toward achieving many of the United Nation's 17 Sustainable Development Goals (SDGs) by following a strategic human capital approach as 11 of the SDGs relate directly (SDGs 3, 4, 5, 8, 9, and 10) and indirectly (SDGs 1, 2, 6, 7, 10, and 16) to human capital. The approach also aligns seamlessly with the aspirations of Agenda 2063 to create "the Africa we want" put in place by the African Union (AU). Effectively achieving the SDGs, and the broad aspirations of Agenda 2063, lies in adopting a bottom-up approach in which problems and challenges are tackled and subsequently linked to the SDGs rather than a top-down approach of taking the SDGs and subsequently looking for the problems they relate to. A strategic human capital approach to development aligns with adopting a bottom-up approach to the SDGs. Beyond its inherent capacity in aiding in the progress toward attaining the SDGs, it is an approach that is also particularly amenable to the peculiar challenges facing African countries that have lagged behind in terms of development for far too long.

Adopting a strategic human capital approach begins with understanding the nature of the current stock of a country's human capital and having a strategic plan for human capital accumulation and development. Consequently, conducting a human capital audit and ensuring continuous tracking of the country's human capital is essential to this approach. For African countries adopting this approach, having education systems that truly result in the formation of human capital, not mere certificates, is paramount. Other specific recommendations that have been discussed in the previous chapters, ideas, lessons from the case studies, and summary contents of each chapter are presented in Table 12.1.

However, this approach to sustainable development is predicated on ensuring that other institutional and socio-cultural factors are in place to facilitate human capital deployment into value-creating productive activities. Research has shown that the fundamentals of sustainable long-term development, physical capital, institutions, and human capital, interact with one another. Institutions, in particular, interact with human capital to produce various outcomes. Consequently, the human capital approach requires good policy frameworks. There also has to be synergy in the public policy and agencies. For example, few countries have agencies for human capital. What is available in most countries are separate agencies for education and health. In reality, it is doubtful that these two agencies work together in many countries. However, successful implementation of the many generic strategies and specific policy recommendations in this book will require the inputs and processes of these two agencies at both national and local levels. Consequently, positive human capital outcomes require public agencies on education and health to collaborate on certain programs that can help improve public outcomes. More generally, sustainable development requires coordinated strategic efforts with programs and policies that are aligned with one another at the national and sub-national levels. This entails an alignment across all public agencies in a country such that the efforts of one public agency or program are not inadvertently sabotaged by those of another public agency.

This approach is anchored on the concept of human capital which is still disputed in some quarters. Since the very beginning, the "human capital" idea has faced many criticisms. Economists, educators, sociologists, and philosophers have all criticized human capital (Tan, 2014). Indeed, "the idea of human capital has been bitterly criticized and sometimes sarcastically referred to as human cattle" (Tan, 2014, p. 431). Many have criticized the idea and theory of human capital asserting that it is unrealistic, emphasizes methodological individualism,[2] has methodological weaknesses, lacks empirical support, and is lacking in morality as it seeks to reduce humans to capital (Bonal, 2016; Bowles & Gintis, 1975; Fine & Milonakis, 2009;

---

[2] The analysis and portrayal of (social) phenomenon as resulting solely from subjective personal motivations.

Marginson, 2019; Tan, 2014; Teixeira, 2014). At the start of the study of the concept in the 1950s and 1960s, even some of those that supported the idea of the human factor contributing to economic growth felt the term human capital was taking the idea too far (Teixeira, 2014). However, despite the many criticisms, human capital has spread far and wide, and its importance continues to grow as well as being studied across many disciplines. No doubt many of the criticisms have some merits. Like all ideas and theories, especially those that relate to humans, their behavior and behavioral outcomes, there are always limitations because they often rest on some explicit and implicit assumptions, and the idea and theory of human capital is not left out. Nevertheless, human capital theory has great utility and continues to serve as a valuable lens through to understand many phenomena. Besides, recent empirical evidences and models of economic growth have shown with reliable proof the impact of human capital on economic growth (Galor & Weil, 2000; Goldin, 2016; Mankiw et al., 1992; Sharipov, 2015).

Criticisms of human capital, and its theoretical foundations in economics, have grown just as its popularity. More than 60 years on from its formal conceptualization and postulations, human capital theory has become very popular. It has been used as the theoretical lens to study a number of issues in many areas, from health, education, and migration to job and firm performance. One foremost area of critic of human capital theory is from some studies of education. Basically, they argue that the idea of human capital reduces education to a mere business activity that is driven solely by profit rather than for other noble ideals such as self-enrichment and human development (Ball, 2010; Marginson, 1997 in Tan, 2014; Nussbaum, 2010). Consequently, education as a component of human capital is seen merely as something that is for trade and business rather than for its other benefits and as a public good that should be provided for by governments. Despite these criticisms, human capital has been used to shape educational policies in many countries (Tan, 2014). In regard to using human capital theory for shaping policies in education, Tan

(2014) writes the following very apt response to the criticisms:

> These critiques give the impression that they have a better alternative model for education policies. But sometimes it is quite noticeable that these criticisms are mostly driven by an ideological zeal with full of resentment just to attack the dominant school of thought with no present alternative at hand. It is true that each criticism is valuable on its own, no matter where it comes from but the ambitious goal, replacing HCT, requires much more than that. There have been huge criticisms but little systematic efforts to understand the origin and impact of HCT (HCT—human capital theory). (p. 437)

The idea of human capital and human capital theory in its diverse utilizations in different disciplines may have its weaknesses which give some value to the criticisms. However, the extant impact of human capital in different disciplines could not have been sustained if it had no utility. Therefore, without delving further into the intricacies of the other arguments and assertions of critics, the approach to sustainable development outlined in this book, while based on many elements of human capital and its theory, does not rely solely on its tenets in economics such as the rational choice theory and methodological individualism. Rather, in addition to the basic tenets of human capital theory in economics, it has drawn from strategic human capital studies in strategy, strategic human resource management, and entrepreneurship, to advance the crucial need for African governments to focus on human capital development as the core means of economic growth and sustainable development. This approach becomes imperative in the face of the continent's youthful population that is set to continue to increase.

One must not, however, ignore the potential difficulties that may be encountered in employing a strategic human capital approach such as the one that has been presented in this book. Two of these potential difficulties are pertinent. One, the approach requires financial resources that may not be easily and readily available which is one reason a multi-stakeholder approach has been recommended alongside in Chap. 11. Two, the long-term nature of returns to human capital may not be easily coupled and aligned with specific political government administrations. Therefore, this may hinder the political will of African governments, particularly those with liberal democratic governments, to make the necessary human capital investments. Research and scholarship into viable ways through which these difficulties may be circumvented will be particularly useful for African countries. Other broad research opportunities are discussed below.

Broad Research Opportunities

The limitations and weaknesses of human capital present inherent opportunities for research. The approach to sustainable development that has been presented in this book also has some inherent limitations. This session presents a brief discussion on four broad areas that have been identified for further research. Research in these areas could provide insights into further understanding human capital, sustainable development, and human capital development in African countries.

More Inter- and Transdisciplinary Studies on Human Capital

The first research opportunity is a call for inter- and transdisciplinary studies on human capital.

It is interesting that although human capital literatures are many and cut across disciplines, there has been very little collaborations between them. Studies of education and education economics discipline, for examples, both deal on the same subject matter—education. Cross-disciplinary studies from both areas on how the notion and theory of human capital "should" or "should not guide" education may serve to enlighten both supporters and critics alike. The society and countries stand to benefit from more robust frameworks on education policies. Perhaps, some of the criticisms and advocates of human capital and its theoretical tenets in other disciplines will benefit from similar cross conversations and interdisciplinary studies. There has

also been scarce conversation between human capital literatures in strategy and human resource management, until quite recently. Health is also a major component of human capital, and there is scarcely any interaction between health-related disciplines and other disciplines in the study of the concept. Therefore, a transdisciplinary perspective may also be necessary to fully understand human capital and the role of its different components, both in research and also in practice.

Human Capital Development, Accumulation, and Deployment in Africa

Second, there is need for more research on the intricacies of how African countries can develop their human capital. African countries, particularly those in Sub-Saharan Africa, have the lowest scores on all global human capital indexes.[3] Despite some criticisms against the components, measures, and methodological issues that have been raised against some of the global indexes that have been used for measuring human capital, the low ranking of African countries on all the indexes cannot be by happenstance. In this regard, coming up with national indexes that measure human capital of sub-national regions is a viable research opportunity. Answers are also needed to some particularly important questions. For example, how can children in rural areas have access to quality education and health-care services in African countries? However, accumulating and developing human capital is not enough. Some of the benefits of human capital result mostly from its deployment in productive activities. Consequently, it is also important for research to explore the effective deployment of human capital. More country-specific studies exploring effective human capital will be highly useful for public policy. In addition, extant research on human capital development for countries has been largely limited to the field of eco-

---

[3]Chapter 5 presents a detailed discussion of the different global human capital indexes.

nomics. Though this has come with great utility, studying human capital development only through the lens of economics has its limitations. Therefore, in studying human capital development in African countries, other disciplines such as strategic management and sociology may need to study human capital development for a more holistic understanding. Understanding human interactions, behaviors, and systems within collective societal organization in the context of countries is important to gleaning insights into human capital development and deployment. Interdisciplinary studies, in particular, may yield useful insights in this regard.

The "Competitive Advantage" and Competitiveness of Countries

Porter (1990, 2000), Delgado et al. (2012), Boikova et al. (2021), Medeiros et al. (2019), Yeganeh (2013), and some others explore the competitiveness of nations. The sustainable development of countries can hardly be discussed without a mention of, and a regard for, national competitiveness. Although countries' sustainable development and national competitiveness are quite distinct, there are considerable overlaps between the two. In one of the positive reviews of Porter (1990), Grant (1991, p. 535) writes that "the breadth and relevance of Porter's analysis have been achieved at the expense of precision and determinancy." The review also states that concepts were poorly defined, theoretical relationships were poorly specified, and empirical data were chosen selectively and were interpreted subjectively (Grant, 1991). However, as also acknowledged by Grant (1991), and by Smit (2010), Huggins and Izushi (2015), and some other reviews, Porter (1990) "shifts focus away from the performance of the firm to the performance of the nation." The weaknesses in Porter (1990) call for research into the competitiveness of countries with the goal of unearthing specific intricacies of how countries can become more competitive, hopefully achieving economic growth and development alongside. The development plight of African countries, and indeed

some other countries, reiterates this call for research into the competitiveness or competitive advantage of countries. Studying the competitiveness of countries will be of interest to students and scholars of international economics, strategy, international business, and similar fields.

Understanding the competitiveness of countries holds promise in aiding the search for country-specific variables and explanations for the successful implementation of national and global strategies. Related to this, extant research has established that human capital resources can be at different levels. It can be at the individual and organizational levels. Firms can have unique capabilities that are rooted in the collective knowledge, skills, and abilities of their employees. As discussed in Chap. 3, countries have also been known to develop unique capabilities. More research is needed on how such unique advantages and collective national human capital and capabilities emerge. Collaborative efforts that cut across traditional boundaries of disciplines could be particularly fruitful in this regard.

Other Approaches to Africa's Sustainable Development

The strategic human capital approach presented in this book is only one approach to sustainable development. It is important to state that adopting this approach does not preclude countries from the concurrent adoption of any other viable approach. The main onus should be on ensuring that the approaches are complementary and not substitutes. In this light, research exploring other approaches to Africa's sustainable development that can be adopted or adapted by countries are urgently needed. In addition, the recommendations and suggestions that have been discussed in this book are not the sole ones that align with a strategic human capital approach to development. Undoubtedly, there are many other recommendations that could potentially align with this approach. Research into additional propositions that align with this approach are fruitful avenues for more insights. More specifically, there is need for research into alternate education and health-care systems especially those that can be based on African indigenous knowledge systems. Research opportunities also exist to explore ways in which the use of modern technologies can propel real human capital formation among the continent's teeming and increasing young and youthful population.

Concluding Remarks

The returns to any human capital development strategy start to become visible in the long term (Diebolt & Hippe, 2019; Hippe, 2020; Acemoglu & Autor, 2012). Unfortunately, unlike investment in physical infrastructure, the yield and dividends of investment in human capital are not readily visible. Any investment in human capital development is a long-term venture and takes time to yield results. The benefits and outcomes of human capital to individuals and organizational groups such as firms, societies, and countries, presented and discussed in Chap. 4, take a while to be visible and felt. However, the results of human capital investments are usually long-lasting and multilayered and impact on various other aspects of the economy and social–cultural lives of people. The results can become more readily visible when other investments are prioritized based on their effect on human capital deployment. African governments will need to spur up the "political" will and long-term commitment to making the necessary financial and nonfinancial investments to developing and accumulating human capital. However, it is still arguable whether democratically elected governments in "liberal" democracies who need to seek reelection every few years can spur up the required political will and commitment for such long-term endeavors. Finally, regardless of the loftiness of any policy recommendation and framework, a lack of will and commitment to implement will almost always impede adoption or adaptation and implementation. African countries need to prioritize investments in human capital and implement policies which encourage investments in human capital by multiple stakeholders, to have the chance of reaping the potential demographic dividends that could result from having a large and growing youthful population.

References

Acemoglu, D., & Autor, D. (2012). What does human capital do? A review of Goldin and Katz's the race between education and technology. *Journal of Economic Literature, 50*(2), 426–463.

Ball, S. (2010). New voices, new knowledges and the new politics of education research: The gathering of a perfect storm? *European Educational Research Journal, 9*(2), 124–137.

Barro, R. J. (1991). Economic growth in a cross section of countries. *The Quarterly Journal of Economics, 106*(2), 407–443.

Becker, G. S. (1962). Investment in human capital: A theoretical analysis. *The Journal of Political Economy, 70*(5), 9–49.

Becker, G. S. (1964). *Human capital: A theoretical and empirical analysis, with special reference to education*. University of Chicago Press.

Becker, G. S. (2007). Health as human capital: Synthesis and extensions. *Oxford Economic Papers, 59*, 379–410.

Bloom, D., & Canning, D. (2003). Health as human capital and its impact on economic performance. *The Geneva Papers on Risk and Insurance, 28*(2), 304–315.

Boikova, T., Zeverte-Rivza, S., Rivza, P., & Rivza, B. (2021). The determinants and effects of competitiveness: The role of digitalization in the European economies. *Sustainability, 13*(21), 11689. https://doi.org/10.3390/su132111689

Bonal, X. (2016). Education, poverty, and the "missing link": The limits of human capital theory as a paradigm for poverty reduction. In K. Mundy, A. Green, B. Lingard, & A. Verger (Eds.), *The handbook of global education policy* (1st ed., pp. 97–110). Wiley.

Bowles, S., & Gintis, H. (1975). The problem with human capital theory – A Marxian critique. *The American Economic Review, 65*(2), 74–82.

Crook, T. R., Todd, S. Y., Combs, J. G., Woehr, D. J., & Ketchen, D. J., Jr. (2011). Does human capital matter? A meta-analysis of the relationship between human capital and firm performance. *Journal of Applied Psychology, 96*(3), 443–456.

Dao, T. B. T., & Khuc, V. Q. (2023). The impact of openness on human capital: A study of countries by the level of development. *Economies, 11*(7), 175. https://doi.org/10.3390/economies11070175

Delgado, M., Ketels, C., Porter, M. E., & Stern, S. (2012). *The determinants of national competitiveness*. NBER Working Paper Series 18249.

Diebolt, C., & Hippe, R. (2019). The long-run impact of human capital on innovation and economic development in the regions of Europe. *Applied Economics,*

51(5), 542–563. https://doi.org/10.1080/00036846.20 18.1495820

Dimov, D. (2017). Towards a qualitative understanding of human capital in entrepreneurship research. *International Journal of Entrepreneurial Behaviour & Research, 23*(2), 210–227. https://doi.org/10.1108/ IJEBR-01-2016-0016

Filmer, D., Gatti, R., Rogers, H., Spatafora, N., & Emrullahu, D. (2021). *Education and health for inclusiveness.* IMF Working Paper WP/21/60.

Fine, B., & Milonakis, D. (2009). *From economics imperialism to freakonomics: The shifting boundaries between economics and other social sciences.* Routledge.

Galor, O., & Weil, D. N. (2000). Population, technology, and growth: From Malthusian stagnation to the demographic transition and beyond. *American Economic Review, 90*(4), 806–828.

Goldin, C. (2001). The human capital century and American leadership: Virtues of the past. *Journal of Economic History, 61*(2), 263–292.

Goldin, C. (2016). Human capital. In C. Diebolt & M. Haupert (Eds.), *Handbook of Cliometrics.* Springer.

Goldin, C., & Katz, L. F. (2008). *The race between education and technology.* The Belknap Press of Harvard University.

Grant, R. M. (1991). Porter's 'competitive advantage of nations': An assessment. *Strategic Management Journal, 12*(7), 535–548.

Hessels, J., Rietveld, C. A., Thurik, A. R., & Van der Zwan, P. (2020). The higher returns to formal education for entrepreneurs versus employees in Australia. *Journal of Business Venturing Insights, 13,* e00148.

Hippe, R. (2020). Human capital in European regions since the French revolution: Lessons for economic and education policies. *Dans Revue D'Economie Politique, 130,* 27–50.

Huggins, R., & Izushi, H. (2015). The competitive advantage of nations: Origins and journey. *Competitiveness Review, 25*(5), 458–470. https://doi.org/10.1108/ CR-06-2015-0044

Mankiw, G., Romer, D., & Weil, D. N. (1992). A contribution to the empirics of economic growth. *Quarterly Journal of Economics, 107,* 407–437.

Marginson, S. (2019). Limitations of human capital theory. *Studies in Higher Education, 44*(2), 287–301. https://doi.org/10.1080/03075079.2017.1359823

Martin, B. C., McNally, J. J., & Kay, M. J. (2013). Examining the formation of human capital in entrepreneurship: Ameta-analysis of entrepreneurship education outcomes. *Journal of Business Venturing, 28*(2), 211–224.

Medeiros, V., Godoi, L. G., & Teixeira, E. C. (2019). Competitiveness and its determinants: A systematic analysis for developing countries. *CEPAL Review, 129,* 2–25.

Mincer, J. (1958). Investment in human capital and the personal income distribution. *Journal of Political Economy, 66,* 281–302.

Nussbaum, M. C. C. (2010). *Not for profits: Why democracy needs the humanities.* Princeton University Press.

Porter, M. E. (1990). *The competitive advantage of nations.* Free Press.

Porter, M. E. (2000). Attitudes, values, beliefs and the microeconomics of prosperity. In L. E. Harrison & S. P. Huntington (Eds.), *Culture matters: How values shape human Progress.* Basic Books.

Schultz, T. W. (1961). Investment in Human Capital. *The American Economic Review, 51*(1), 1–17.

Sharipov, I. (2015). Contemporary economic growth models and theories: A literature review. *CES Working Papers, ISSN 2067-7693, Alexandru Ioan Cuza University of Iasi, Centre for European Studies, Iasi, 7*(3), 759–773.

Smit, A. J. (2010). The competitive advantage of nations: Is Porter's diamond framework a new theory that explains the international competitiveness of countries? *Southern African Business Review, 14*(1), 105–130.

Smith, A. ([1776] 2009). An inquiry into the nature and the causes of the wealth of nations. In C. Muir & D. Widger (Eds.), *Project Gutenberg.*

Tan, E. (2014). Human capital theory: A holistic criticism. *Review of Educational Research, 84*(3), 411–445. https://doi.org/10.3102/0034654314532696

Teixeira, P. N. (2014). Gary Becker's early work on human capital – Collaborations and distinctiveness. *IZA Journal of Labor Economics, 3*(1), 12. https://doi.org/10.1186/s40172-014-0012-2

Unger, J. M., Rauch, A., Frese, M., & Rosenbusch, N. (2011). Human capital and entrepreneurial success: A meta-analytical review. *Journal of Business Venturing, 26,* 341–358.

Wilson, R. A., & Briscoe, G. (2004). The impact of human capital on economic growth: A review. In P. Descy & M. Tessaring (Eds.), *Impact of education and training: Third report on vocational training research in Europe: Background report* (Cedefop Reference series, 54). Office for Official Publications of the European Communities.

Yeganeh, H. (2013). An investigation into the cultural and religious determinants of national competitiveness. *Competitiveness Review, 23*(1), 23–40. https://doi.org/10.1108/10595421311296605

Index

E. Aziegbe-Esho, *On the Sustainable Development of African Countries*, Sustainable Development Goals Series, https://doi.org/10.1007/978-3-031-81124-1

The manufacturer's authorised representative in the EU is Springer
Nature Customer Service Centre GmbH, Europaplatz 3, 69115 Heidelberg,
Germany. If you have any concerns regarding our products, please
contact ProductSafety@springernature.com

Printed and bound by CPI Group (UK) Ltd, Croydon, CR0 4YY
28/05/2025
01885623-0001